Fixings, fasteners and adhesives

PAUL MARSH

Construction Press

LONDON AND NEW YORK

Construction Press
An imprint of:
Longman Group Limited
Longman House, Burnt Mill, Harlow
Essex CM20 2JE, England
Associated companies throughout the world

*Published in the United States of America
by Longman Inc., New York*

First published 1984

British Library Cataloguing in Publication Data
Marsh, Paul, 1931–
  Fixings, fasteners and adhesives. — (Site
  practice series)
  1. Fasteners
  I. Title   II. Series
  693   TH153

621·88

MAR

  ISBN 0-86095-035-2

Library of Congress Cataloging in Publication Data
Marsh, Paul Hugh.
  Fixings, fasteners, and adhesives.

  (Site practice series)
  Includes index.
  1. Joints (Engineering)   2. Fasteners.
3. Adhesives.   I. Title.   II. Series
TA660.J64M37   1984   621.8′8   83-7465
  ISBN 0-80095-035-2
Set in 10/12 pt Lintron 202 Bembo
Printed in Hong Kong by
Commonwealth Printing Press Ltd

Contents

Preface

As the number and importance of fixings and fasteners have increased in recent years, so has the need to ensure their correct use on site. Faced with a bewildering multitude of devices and adhesives, the first problem is, which fixing to use; the second, how to use it so that it achieves its maximum effect.

This book is intended to help the practical man with both of these problems. It will not attempt to advise on the design of heavy duty structural connections – this work is the domain of the engineer. It will, however, give advice on the installation of the device specified by the engineer and, where a choice exists, explain the different types of fixings and fasteners available, and which should be used where.

Fixings will be considered within five major categories – the first four consisting of *mechanical fixings* (all fixings except adhesives), the fifth consisting of adhesives (*chemical fixings*).

Mechanical fixings are divided into four categories depending on the type of base material into which they are making a fixing (mass walling, timber, metal or cellular materials). A more detailed explanation of these divisions will appear in the next chapter when we discuss types of fixings. It is sufficient here to say that the reason for this choice of categories is because the character of the base material exerts considerable influence over the type of fixing which can be used, and therefore a classification by base material makes good sense.

Each category of fixings will be dealt with in a similar way. The chapter will start with a brief description of the problem of making fixings to the particular base material; then each fixing device or fastener will be dealt with individually. It will be described, its uses or applications discussed, and the method of its setting or installation given. These *setting instructions* represent the minimum amount of information necessary in order to use the fixing correctly. For quick, practical assistance, this is the only part of the chapter that need be read, along with the general advice contained at the beginning of each chapter and a glance at the appropriate illustrations. For those who are interested in finding out more, there is a section at the end of each chapter giving additional background information concerning each category of fixing. This will help to give a more broad understanding of the fixing being considered and may cast extra light on some of the advice given in the previous parts of the chapter.

In Appendix 1 appears a schedule of proprietary devices divided into the same categories, along with manufacturers' names and addresses. Manufacturers will always be pleased to give advice on the use of their products.

One word of warning: as the complexity of fixing devices grows day by day, more sophisticated devices are being added to the product ranges. The advice given in this book is applicable to the type of device concerned, but may not be wholly correct for newer or more specialized devices. It is always advisable to read and follow the manufacturer's instructions when using a new mechanical fixing device or adhesive of which you have no previous experience.

1

The importance of fixings

The importance of fixings in the building industry has grown immeasurably over the last fifty years. As construction materials have generally increased in efficiency and strength, so structures have tended to become lighter, made up of materials with a higher strength/weight ratio. However, as a direct result of their lightness, structures have become more vulnerable to the effects of wind pressure, accidental impact damage and similar misfortunes. No longer can the sheer weight of a building, combined with the low-grade adhesion that mortar provided, be considered sufficient to hold the parts of the structure together. High-strength mechanical and adhesive fixings have had to come to the rescue, making good the structural shortfall that reduced weight brought in its wake.

What is more, as structural components have become lighter, they have also tended to become smaller, presenting less area of contact between the parts of the joint for whatever fixings may be necessary to hold them together. So twelve nails in a joint for safety's sake where practically four would do, is no longer physically possible — even if it were economically permissible today to use more fixings than are really necessary. For this reason alone fixings have often to be fewer in number, and therefore each has to be of greater strength and of thoroughly predictable performance.

All this has resulted in the growth of a new breed of high-performance fixings with assured engineering achievement. Also, as structural elements have become thinner, the character of fixings has had to change, leading to ever more sophisticated fasteners and adhesives which can make fixings to thin or cellular materials, like honeycomb core partitions and doors — fixings which would not have been possible a few years ago.

But it is not only in the design of fasteners and adhesives where the development pressure is on. The strength of the device or glue alone is not enough to determine the strength of the joint. The method and skill of its application or use, too, is a vital factor in the efficiency of the fixing.

It is for this reason that this book will concentrate on the practical techniques to be used in the installation of devices or the use of adhesives so that they can achieve their maximum efficiency. A clumsily-placed fastener, or an adhesive joint closed after the glue has exceeded its open time, can undermine the effectiveness of the joint and (in the case of medium- or heavy-duty fixings) make nonsense of the engineering design of the structure, which will have been calculated in many cases on the expected performance of the fixing or adhesive used with a particular material.

Finally, it is important to remember that even the most expensive fixing forms a very small part of the total cost of a building. A failure of one of these apparently minor devices, however, can cause damage (to life and property) out of all proportion to its cost. Fixings, their design and application, are vital to the performance of the building.

2

Types of fixings

Fixings in the building industry tend to be made between a *base* (usually the most substantial of two elements jointed together and often the first one to be constructed) and a *subsidiary component* (the less substantial of the two elements). Typical examples are:

(a) a fixing between the structural frame of a building (the base) and its claddings (the subsidiary component);

(b) or between a loadbearing wall (the base) and loadbearing components, such as floor joists, wall plates, roof trusses, bearers, etc. or special facing material (the component);

(c) or between a non–loadbearing cellular partition (the base) and a shelf bracket (the component);

(d) or between a wall (the base) and a skirting board (the component);

(e) or between a screeded subfloor (the base) and the floor finish (the component).

Clearly the strength required of the fixings in the above examples varies considerably; from the heavy-duty fixing with a reliable structural performance, through to the light-duty fixing for a minor component, such as a skirting board or a floor finish, which carries little or no load, except possibly the weight of the component. The strength of the fixing will always be affected by the strength of the base and of the component.

It is sometimes difficult to decide which of the two parts connected by a fastener is the base and which is the component. Usually the 'heavier' of the two materials is the base (the structural frame, rather than the cladding); but this is not always the case. Sometimes the two materials are identical in 'weight' (e.g. two similar-sized pieces of timber). This is usually the case where both elements being connected are parts of the same structural unit – members of a timber truss, or leaves of a cavity wall. In this case,

3

the question of which of the two elements is the base is unimportant; but this situation is relatively uncommon and so throughout this book we shall refer to the *base* and the *subsidiary component*.

Fixings fall into two broad categories – *mechanical fixings* which involve a metal (or sometimes plastic) nail, bolt, screw, dowel or socket to connect the component to the base, and *chemical fixings* which consist primarily of adhesives. Two types of chemical fixing – the chemical anchor and the injection anchor – although they both rely on the setting of an adhesive, are more like mechanical fixings and therefore will be included in the appropriate mechanical fixing section of this book. All other adhesives will be examined together in Chapter 7.

Mechanical fixings, because the base usually dictates the design of the fixing, will be grouped according to the type of base material in which they are used:

(a) mass walling (brick, stone, concrete blocks and concrete structure);
(b) timber and timber-based products;
(c) metal – heavy or lightweight;
(d) cellular materials (hollow core doors, panels and partitions) and thin wall materials (plasterboard, plastic sheets, etc.).

The purpose of fixings

Fixings are necessary to hold the parts of a joint together. They usually have to be strong enough to carry the weight of the subsidiary component, together with any load that it might have to carry (i.e. its self-weight and its imposed loadings). These are said to be *loadbearing* fixings.

Sometimes a fixing is merely intended to locate the subsidiary component, while its self-weight and its imposed loadings are carried on other fixings, or on a ledge formed in the base material. These fixings are called *restraint fixings*.

Usually a fixing is subjected to forces which try to tear the joint apart. These can operate in *tension*, resulting from a force acting parallel to the axis of a mechanical fixing and in a direction away from the base material. This type of force tries to break the fixing (whether it be the shank of a screw or the film of adhesive), or to pull the fixing out of, or through, the base or the component. In connection with tension forces we therefore refer to *pull-out failure* or *pull-through failure* (Fig. 2.1).

4

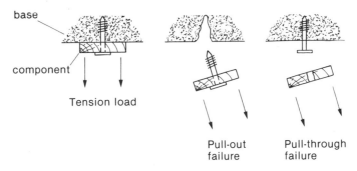

Tension load

Pull-out failure

Pull-through failure

Fig. 2.1 Failures due to tension loading

Joints are often subjected to *shear* forces. These forces act at right-angles to the axis of the fixing. This type of force tries to distort or break the shank of the mechanical fixing, or tear the fixing across the component. This failure is referred to as *shear failure* (Fig. 2.2).

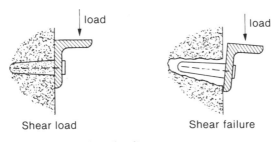

Shear load

Shear failure

Fig. 2.2 Failure due to shear loading

Joints can be subjected to *compression* forces – forces acting parallel to the axis of a mechanical fixing and towards the base material. These tend to push the component into position against the base and hold it there. Compression is rarely the cause of fixing failure. It usually assists the fixing to carry out its task (Fig. 2.3).

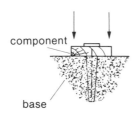

Fig. 2.3 Compression on a joint

5

How fixings work

Fixings work in one of three ways.

Type 1. Squeezing the component to the base

This is called a *through-fixing*. The device passes through both component and base and clamps the two together, usually as a result of tightening a screw (Fig. 2.4).

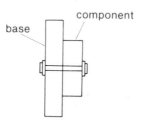

Fig. 2.4 Through-fixing

A through-fixing's strength depends firstly on the materials being secured together being strong enough not to allow the head of the fixing (or its washers or nut) to pull through the material and secondly on the strength of the shank of the device to resist fracture or distortion due to shear loading. Examples of this type of device include bolts, rivets, lightweight toggle fasteners and timber connectors (split-ring or toothed connectors). Generally this type of fixing generates considerable friction between the contact faces of the base and component and makes particularly strong shear joints.

Type 2. Making an anchorage in the base

The base material is penetrated by the anchorage fixing, but only enough to allow it to hold fast and make a secure fixing to support the load of the subsidiary component. There are three kinds of anchorage fixing. Those which depend on:

(a) making an anchorage in a pre-drilled hole (a *drilled-for fixing*);
(b) deforming the base material by driving or firing the fixing into the material, forcing it to be compressed into itself (a *base-deforming fixing*);

6

(c) building-in or casting-in a fixing into the base material.

All these devices achieve their pull-out strength by the friction of the body of the fastener against the base material (Fig. 2.5).

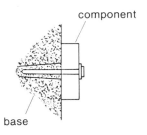

Fig. 2.5 Anchorage fixing

Type 2(a) devices include expanding anchors, plugs, chemical anchors and split-ring connectors.

Type 2(b) devices include nails, nail plates, screws and fired fixings actuated either by an explosive charge or compressed air.

Type 2(c) devices include joist hangers, sockets, corbels, wall ties and cast-in channels.

Anchorage devices are effective when pull-out or shear strength is required.

Type 3. Adhesion

The component is stuck to the face of the base material with no prior treatment to either material, other than smoothing their surfaces and spreading the adhesive. Glues tend to perform most effectively in shear rather than in tension (Fig. 2.6).

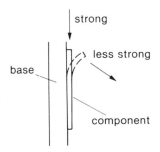

Fig. 2.6 Adhesive joint

7

In addition to the categories above there is a range of accessories intended to be used with mainly type (2) anchorage devices to provide loadbearing or restraint fixings for subsidiary components. These will be dealt with as accessories in the appropriate sections of this book.

How fixings are chosen

The choice of fixing type will depend firstly on the character and size of the base material, secondly on the character and size of the subsidiary component and lastly on the type, size and direction of the load which it is going to have to withstand.

The size and character of the base material dictates the type of fixing to be chosen, hence the division of mechanical fixings into base material groups in this book. The base's character alone will decide whether an expanding anchor can be used or if a particular type of nail or screw fixing is feasible.

The strength of the subsidiary component, too, needs to be considered if pull-through failure is to be avoided with certain types of fixing. It should be remembered that a fixing's strength is directly related to the strength of the base and of the component; and a joint can only be as strong as the materials joined.

The aim of a fixing is to achieve a secure joint which, in the case of a structural or high-performance fixing, must perform to a known standard. To this end fixings manufacturers produce load tables for their heavy-duty fixings used with base materials of specified known strength.

There is a number of secondary factors that need to be considered when choosing a fastener. Will the joint be subjected to thermal, moisture or other structural movement which may occur in the building – and if so, can the fixing accommodate the movement? Is the fixing made of a material that is compatible with the base and component materials, and are the base and component materials themselves compatible? This is an important point. If there is incompatibility of material, corrosion could result, leading to early breakdown of the joint. A typical case where care is needed is where a preservative has been used to treat the timber members of a truss. Some preservatives could cause the nail plates used in the truss manufacture to corrode. If there is any doubt about the side-effects of such treatments, the manufacturer should be consulted.

Corrosion

The problems of corrosion are particularly important when dealing with metal fixing devices and these problems will be referred to throughout the book when appropriate. Generally, corrosion falls into two categories: galvanic corrosion and oxidation (or rusting).

Galvanic corrosion

This is an effect which comes about when two metals are in contact and can result in one metal being eaten away. It is often called contact corrosion, or bi-metallic, or electrolytic attack. Some metals can be used together; others cannot. It all depends on the difference in voltage potential between the two metals. Table 2.1 (taken from information supplied by Harris and Edgar Ltd) indicates which metals can and cannot be used in contact. Where it is essential that incompatible metals are used together, the one must be shielded from direct contact with the other by a neoprene sleeve or gasket.

Oxidation

Steel or malleable iron fixings, because they are subject to oxidation in damp conditions, should not be used unless they have been zinc coated. There are three processes available:

1. *Hot-dip galvanizing*. The dipping of the element in molten zinc gives a good, but uneven, protection and may entail re-drilling or re-threading small holes and threads, with consequent loss of protection.
2. *Zinc plating*. This is an electro-plating process which gives a uniform and ductile protective coating. Re-drilling and re-threading is not usually necessary.
3. *Sherardizing*. This is a process involving the diffusion of hot zinc dust on to the element, giving a thin, uniform coating. It is often not recommended for threaded elements as there is a tendency for the coating to build up on the threads.

No zinc-coating process is completely without the danger of minor defects or damage to the coating which could be the starting point for corrosion. It is frequently recommended that any heavy-duty fixing in locations with little or no access (where

Table 2.1 Degree of galvanic corrosion hazard in bi-metallic contacts (derived from information provided by Harris and Edgar Ltd)

	Copper	Phosphor bronze	Aluminium bronze	Stainless steel	Mild steel	Manganese bronze	Aluminium	Cast iron
Copper	S	S	S	D	X	D	X	X
Phosphor bronze	S	S	S	D	X	D	X	X
Aluminium bronze	S	S	S	D	X	D	X	X
Stainless steel	D	D	D	S	X	D	X	X
Mild steel	X	X	X	X	S	X	X	S
Manganese bronze	D	D	D	D	X	S	X	X
Aluminium	X	X	X	X	X	X	S	X
Cast iron	X	X	X	X	S	X	X	S

S = safe combination
D = safe combination in dry conditions only
X = combination never to be used

corrosion could take place without detection) should be made of non-ferrous metal.

Certainly stainless steel or other non-ferrous devices should always be used in positions of extreme hazard (chemical plants, sewage works or marine locations) or where conditions of prolonged dampness are expected.

3

Mechanical fixings in mass walling bases

All fixings made into mass walling bases have their performance influenced by the strength of the base (Table 3.1).

Density and compressive strength are the two characteristics that most affect fixing performance. Dense masonry generally gives the more effective support for fixings, but it is usually more difficult to fix to a dense masonry base because of the difficulty of drilling or firing into it. Also the accuracy of holes drilled into dense masonry is not always as good as it should be, dramatically affecting the performance of the fixing. Low-density masonry, on the other hand, gives little support to fixings like expanding anchors and, once more, accurate drilling can be a problem. In the case of aerated concrete – the masonry with the lowest density – expanding anchors should not be used.

A few guide figures to masonry materials' vital statistics are given in Table 3.2.

Manufacturers produce performance data for their fixings when applied to base materials of particular compressive strength. These are usually the result of tests performed in accordance with BS 5080: Part 1: *Method of Test for Structural Fixings in Concrete and Masonry* and the recommendations of the Construction Fixing Association. If the base material does not compare with that specified in the manufacturer's data, the company should be consulted and, maybe, further site tests should be carried out, if the performance of the fixing is structurally important.

It must be remembered that no fixing into masonry can be stronger than the masonry itself and generally the deeper the fixing, the greater the volume of masonry resisting its withdrawal and therefore the stronger the fixing (Fig. 3.1).

Safety factors Because, even under controlled test conditions, fixing performance can vary, it is normal to determine working loads by applying a safety factor to a fixing's *ultimate load* (the maximum load it can carry before failure) or alternatively to what is called its *first movement load* (the point at which the fixing is first observed to move 0.2 mm).

For tensile loading, the working load is usually taken as either $\frac{\text{first movement load}}{2}$ or $\frac{\text{ultimate load}}{5}$.

Shear loads are taken as 75 per cent of safe tensile load and tensile shock loading as half the safe sustained tensile load.

Section A: Drilled-for fixings – type 2(a)

The technique of drilling

Before dealing with the fasteners in this category, it would seem wise, as they all depend on the effective and accurate drilling of holes in the mass walling base, to examine drilling techniques and the type of equipment used.

Drilling in masonry is not a cutting operation, as it is in timber or metal; it is a pulverizing and scraping process, involving abrasion of the drill bit. Masonry bits, therefore, have hard carbide tips to give maximum wear.

There are six methods of forming holes in mass walling:

1. By hand, using a percussion tool

This is a slow and laborious method, particularly in hard walling. It can, however, result in accurate holes, particularly for the smaller plug-type fixing.

2. Hand rotary drill

Once more a slow process in hard walling, particularly when drilling larger holes. In softer walling, like common brickwork or lightweight concrete, it produces accurate holes.

3. Electric rotary drill

This is an accurate method of drilling holes, but it can be very slow when the hole diameter exceeds 16 mm in hard walling. It is recommended for softer mass walling bases, such as common brickwork or lightweight concrete.

4. Electric rotary/ impact tool

An accurate method of drilling holes in hard walling materials, but its

Table 3.1 Mechanical fixings in mass walling bases: which fixing to use, performance, base requirements

Fixing type	Duty			Base				
	Light	Medium	Heavy	Dense concrete	L/w concrete	L/w concrete block	Aerated concrete	Aerated concrete block
Plugs; fibrous/plastic	√			√	√	√	√	√
Hammer-set plug	√			√	√	√		
Plugging compound	√			√	√	√	√	√
Expanding anchor		√	√	√	√	√		
Self-drill anchor		√	√	√	√	√		
Ceiling suspension anchor		√		√				
Insulation fastener	√			√	√	√	√	√
Cavity wall repair (mech.)		√	√			√		
Chemical anchor		√	√	√	√	√		
Cavity wall repair (chem.)						√		
Injection anchor	√	√					√	√
Screw-in anchor	√	√		√	√	√		
Masonry nail	√			√	√	√		
Powder-actuated pin		√	√	√	√	√		
Pneumatic-actuated pin		√	√	√	√	√		
Hammer-in anchor	√						√	√
Joist hanger		√				√		√
Anchor strap		√				√		√
Cavity wall tie		√				√		√
Cramp/dowel		√		√		√		√
Cast-in plug	√	√		√	√			
Cast-in socket		√	√	√	√			
Cast-in channel		√	√	√	√			
Cast-in corbel		√	√	√	√			

Non-fines concrete	Brick-work: solid	Brick-work: cellular	Clay block: cellular	Stone: hard	Stone: soft	Services needed	
						Electricity	Compressed air
	√	√		√	√	√	
	√	√		√	√	√	
√	√	√				√	
√	√			√	√	√	
√				√	√	√	
						√	
	√			√	√	√	
	√			√	√	√	
√	√			√	√	√	
	√			√	√	√	
		√	√			√	
	√				√	√	
	√				√		
	√				√		
	√				√		√
	√	√					
	√	√					
	√	√	√	√	√		
				√	√		

Table 3.2

Material	Density (kg/m³)	Compressive strength (N/mm²)
Structural concrete	2400	25 – 40
Staffordshire blue brick or similar	2000	48.4
Lightweight concrete	2000	15 – 30
Common brickwork	1760	20.0
Aerated concrete	700	2.8 – 5.5

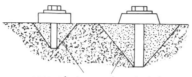

area of stressed material
resisting withdrawal

Fig. 3.1 Effect of the depth of a fixing

effectiveness falls off when the hole diameter exceeds 16 mm. It is recommended for most mass walling bases, except those which are extremely hard.

5. Electric hammer drill

This is a quick method of forming holes up to 24 mm diameter with reasonable accuracy in dense concrete, hard brickwork and hard stone.

6. Pneumatic hammer drill

This is the fastest method, but hole accuracy and shape are not always as good as they might be. This method is recommended for large-diameter holes for heavy-duty civil engineering fixings.

Drilling speed Masonry drills produce their best performance when used at medium or slow drilling speeds (400 to 600 rpm, fully loaded). High speeds result in overheating and consequent premature drill wear. Lower speeds, on the other hand,

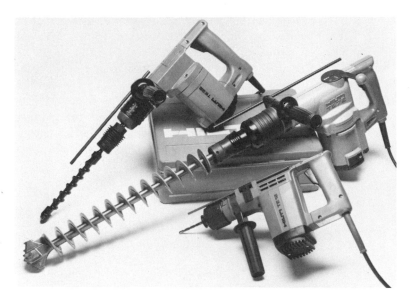

Fig. 3.2 The Hilti range of electro-pneumatic drilling machines

generate less heat which can be dissipated up the drill shank (Fig. 3.2).

De-speed adaptors are available to enable high-speed drills to operate at lower speeds. They also permit high-speed drills to make much larger holes.

Where a walling base is faced with a delicate material, such as glass or glazed tile, it is wise to use a Durium-tip drill at slow speeds.

Impact adaptor There are adaptors available which convert rotary drills into impact tools. These usually only permit the use of the smaller sizes of drill bit.

Points to remember

Whatever method of drilling is used, certain points should be observed:

1. The correct combination of rotary drill or hammer, drill bit or tool should always be used. The size of drill bit must always be that recommended for the particular size of fixing. This is essential if the fixing is to achieve its full performance.
2. Every rotary drill or hammer is limited in its performance

17

by the largest drill bit diameter that is recommended by the manufacturer for use in a specified base material. These recommendations should always be followed.

3. The procedures given in the manufacturer's operating instructions for the tool should always be observed, including work method, cleaning and servicing.

4. After each hole has been drilled (and before the fixing is placed in position) the hole should be cleared of cutting dust, preferably by using a blow-out bulb. The efficiency of the fixing can depend on this (Fig. 3.3).

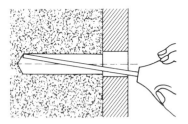

Fig. 3.3 Using a blow-out bulb

5. Finally (and most importantly) safety glasses should always be worn when drilling to prevent dust and chippings flying into the eyes. These can usually be worn over any normal glasses and are made of shatter-resistant polycarbonate.

A1 Expanding anchors and plugs

Fixing devices contained in this section include:

(a) plugs (fibrous or plastic);
(b) hammer-set plastic plugs;
(c) asbestos-based plugging compound;
(d) expanding anchors of three types: projecting bolt type – loose bolt type – socket type;
(e) self-drilling expanding anchors;
(f) expanding anchor: (non-friction type).
(g) specialized anchor devices: including ceiling suspension anchors – insulation fasteners – cavity wall repair anchors;

The common factor between all fixings in this category is that they make an anchorage in the base material by their expansion within a drilled hole in the base. (The self-drilling expanding anchor is an exception to this rule, because it drills its own hole.) The expansion of the fixing presses on the sides of the drilling, building up friction which gives the device its holding power.

Advice on drilling holes in masonry bases is given in the previous section. The drilling of accurate holes of the correct diameter and depth is essential to the performance of the device.

The strength of these fixings varies greatly from the lightweight fixings made by plugs and insulation fasteners to the heavyweight structural performance demanded of larger expanding anchors. In all types of fixing, however, the care with which it is set in the base is critical to its performance.

Further information on the use of anchors in various types of mass walling base, their positioning and other matters affecting their performance are dealt with in the background section on page 63.

Plugs

Description These are light–duty, screwable inserts, placed in mass walling to receive woodscrews or coachbolts used to secure the subsidiary component. In the past they have been made of wood; today they can be made of a natural fibre (as in the Fibre Rawlplug) or plastic (polypropylene or nylon). In each case the action of driving the screw or coachbolt expands and distorts the plug against the sides of the drilled hole, filling up any surface irregularities of the drilling. Once a plug is placed, it cannot usually be easily removed, but plugs to which fixings have already been made can often receive a later screw fixing.

Applications Plugs are used for making relatively light-duty fixings to any mass walling material and are particularly applicable for use in soft building materials, such as aerated or lightweight concrete, for which some specially profiled plugs have been developed. Some of these are of the hammer–home variety: some like the Fischer Twist Lock anchor are recommended for aerated concrete; others like the Rawlplug Hammer Screw are not (Fig. 3.4).

The performance of all plug fixings depends on the drilled hole being of the correct size for the plug and the plug being of the correct size for the screw. In the case of irregular or ragged-shaped

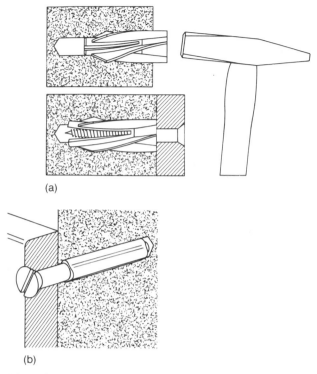

(a)

(b)

Fig. 3.4 (a) Fischer Twist Lock anchor (b) Rawlplug Hammer
Screw

holes (or holes in poor-quality masonry) an asbestos-based compound (Rawlplastic) can be used instead of a preformed plug.

Types of plug

Fibrous plugs: To receive woodscrews sizes from 6 to 20 or
                 coachscrews sizes from 6 to 8 mm.

Plastic plugs: To receive woodscrews sizes 4 to 20. In the
                 case of plastic plugs one size of plug is recom-
                 mended for several screw sizes, so that some
                 manufacturers are able to cover the range of
                 screw sizes in 6 or 7 plug diameters.

Hammer-set
plastic plugs
Asbestos-based
compound: To receive woodscrews sizes from 4 to 20.

Note: Some special plastic plugs are manufactured for use in lightweight concrete bases only. These are usually only available in a few sizes in the middle of the upper range of screw sizes (see Fig. 3.4(a)).

Setting instructions

Fibrous plugs

1. Drill hole of the recommended diameter and to a depth equal to the length of the plug.
2. Insert the screw one or two turns into the plug and use the screw to push the plug into the hole.
3. Turn the screw into the plug to the extent of the thread only.
4. Withdraw the screw leaving the plug set in the base.
5. Place the subsidiary component in position and replace the screw, driving until tight.

Note: The plug should always be the same length as the thread on the screw and the shank of the screw should preferably not enter the plug if damage is to be avoided. The plug should be set completely into the walling, below the surface of the plaster.

Plastic plugs

1. Drill hole of the recommended diameter.
2. Insert the plug so that its flange is on the surface of the mass walling base (or its plaster).
3. Pass the screw through the subsidiary component into the plug and drive the screw home.

Note: This is the procedure for plastic plugs with flanged tops. Other manufacturers produce non-flanged plugs which can be inserted on the end of the screw, through the subsidiary component, into the drilled hole, and then the screw can be tightened.

Hammer-set plastic plugs
(These are push-through fixings in which the plastic plug is already fixed on the point of the screw or nail; see (b) in Fig. 3.4.)

1. Select a device longer than the thickness of the subsidiary component plus the plaster. This is the length of the plug

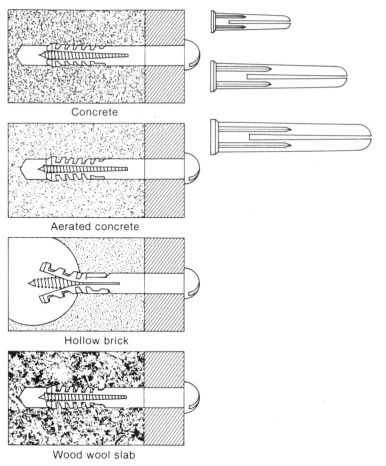

Concrete

Aerated concrete

Hollow brick

Wood wool slab

Fig. 3.5 Fischer Wallplug Type S and Split plastic plug range

that will be set in the walling. In heavier fixings this measurement should be at least 25 mm.

2. Drill hole of the recommended diameter and 5 mm deeper than the fixing.

3. Position secondary component over the hole and insert fixing and hammer the screw (or nail) home.

4. A screwdriver can be used as a setting tool where the hammer cannot reach the head of the fixing. In the case of screw fixings, a final tightening can be made by screwing.

Note: Those devices using a drive screw, rather than a nail, can be easily unscrewed at some future date if the component needs to be removed. Refixing is possible by replacing the screw.

Asbestos-based compound
1. Drill a hole of a diameter no less than that of the screw.
2. Take a small quantity of the compound, immerse it quickly in water and squeeze and roll it into the form of a plug.
3. Ram the plug into the hole with a tool provided with the compound and, using the same tool, make a small lead hole in the surface of the plug when the drilling is completely filled.
4. Pass the screw through the subsidiary component and drive into the plug, being careful not to over-drive.

Note: Compounds contain asbestos and therefore should be handled with care. Always keep the container closed. Avoid inhaling the fibres. Dampen the product immediately it is removed from the container and always thoroughly wash your hands after using the compound.

Asbestos-based plugging compound can be used when the hole to receive the screw fixing is irregular or over-sized for the correct size of pre-formed plug. Thus drilling mistakes can be rectified.

Expanding anchors

Description Generally these represent the more heavy-duty drilled-for-fixings in mass walling. They are normally manufactured of steel (galvanized or zinc-plated), stainless steel or aluminium bronze, although there are heavy glass-reinforced nylon products on the market. They all depend for their holding power on the expansion of a part of the device within the drilled hole. This compresses the surrounding mass walling and produces a high frictional resistance to pull-out. The expanding section of the device can consist of metal or PVC sleeves, or a metal shell, depending on the pattern of the device (Fig. 3.6).

Applications Expanding anchors are used to make medium- to heavy-duty fixings with a predictable performance into a variety of walling materials, including dense concrete, no-fines concrete, lightweight concrete with a dense texture, stone

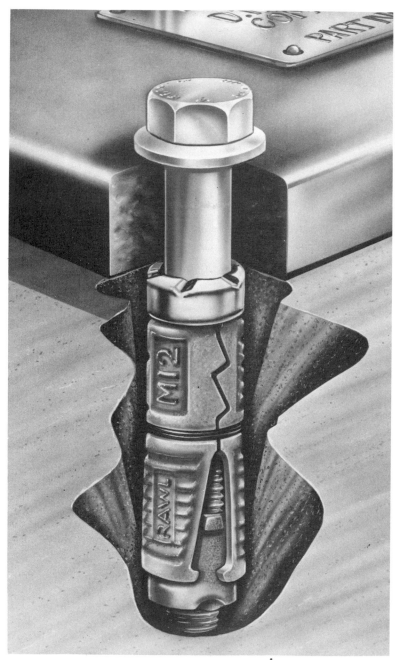

Fig. 3.6 Rawlplug expanding anchor (artist's impression)

and brickwork made up of solid bricks (without large internal voids). In the case of brickwork or blockwork, anchors should be set as centrally as possible in the bricks or blocks. In the case of bricks which are exceedingly hard and brittle, anchors can be placed in the mortar joints (but their pull-out strength may be affected).

Expanding anchors should not be used in aerated concrete – only lightweight fixings can be made into this material using injection anchors (see sect. A3, p. 39) or specially developed plastic plugs (Hilti H6 and FD anchors) or the Rawlnut (see Ch. 6).

Self-drilling anchors should not be used with any brick or clay block base.

An anchor's performance will vary with the strength of the base into which it is set and its position in the base in relation to its edges. Because of the expanding action of these devices, they can cause cracking or spalling if placed too close to the edges of the base. Simple rules of thumb that can safely be applied are: when the load is at right angles to the edge of the base, the minimum distance between the centre of the anchor and the edge of the base should be 2.5 to 4 times the anchor's length; when the load is parallel to the edge, 1.5 to 3 times the anchor's length. If anchors are placed too close together (less than 2 to 3.5 times the length of the anchor) their performance can also suffer. These and other matters affecting the performance of expanding anchors will be dealt with in the background part of this chapter.

As the outside diameter and length of an anchor increases (all other factors being equal) so its load capability increases. This, however, is dependent upon the correct drilling of the hole.

Over-deep holes can reduce fixing performance unless the device is the type which can be sleeved adequately. Holes of too large a diameter reduce the pull-out strength of the anchor because its expansion is not able to compress the base sufficiently. The anchor needs to be fully expanded in the right-sized hole to achieve its best performance.

Types of anchor

Projecting bolt type: These devices have a captive bolt or stud. Some are push-through fasteners (in other words the subsidiary component can be positioned before the anchor is placed in the drilling); some need to be set and the subsidiary component positioned over its projecting stud. Sizes vary from bolt diameters of 5 to 24 mm. Thicknesses of subsidiary components fixed can be up to 115 mm (Fig. 3.7).

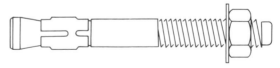

Fig. 3.7 Projecting bolt type of anchor

Loose bolt type: This anchor is set before the subsidiary component is positioned. The size range is approximately the same as the projecting bolt type of anchor (Fig. 3.8).

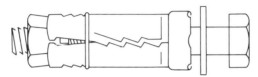

Fig. 3.8 Loose bolt type of anchor

Socket type: This is a threaded socket anchored in the base (much like a heavy-duty plug) prepared to receive a bolt fixing. Sizes suit bolts with diameters of 6 to 24 mm (Fig. 3.9).

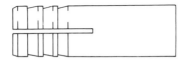

Fig. 3.9 Socket type of anchor

There is a variety of adaptors, particularly for use with loose bolt or socket types of anchor which vary the head style of the fixing (such as the hook-and-eye adaptor) or allow the anchor to be used in alternative situations (such as extended studding and metal or plastic collar adaptors or sleeves), (Fig. 3.10).

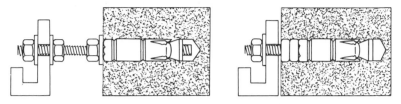

Fig. 3.10 (a) Extended studding (b) plastic collar adaptors

Setting instructions

Projecting bolt type (push-through version)

1. Drill a hole of the recommended depth and diameter. Often in this type of anchor the diameter of the drilling is the same as that of the device.
2. Insert the anchor through the subsidiary component with the nut and washer on the end of the stud.
3. Tighten the nut to the recommended torque. This expands the anchor and secures the component.

Note: The fact that the subsidiary component can be in place during the whole fixing operation is particularly useful when the component is heavy or awkward to move. In the case of some lighter devices used to fix door linings and window frames, the subsidiary component can be positioned and drilled through at the same time as the base is drilled. Some of these devices, known as *frame anchors*, have screw (rather than bolt) heads and are specially designed for stand-off fastenings where there is a gap between the back of the subsidiary component and the base.

Projecting-bolt type (non-push-through version)

1. Drill a hole of the recommended diameter and depth.
2. Insert the anchor.
3. Position the subsidiary component over the threaded studding of the anchor.
4. Apply the nut and washer and tighten to the recommended torque to expand the anchor and secure the component (Fig. 3.11).

Loose bolt type

1. Drill a hole of the recommended diameter and depth.

27

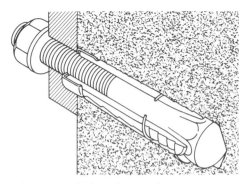

Fig. 3.11   Projecting bolt anchor in place

2. Insert the body of the anchor.
3. Position the subsidiary component over the anchor.
4. Pass the bolt through the subsidiary component into the body of the anchor and tighten to the recommended torque to expand the anchor and secure the component (Fig. 3.12).

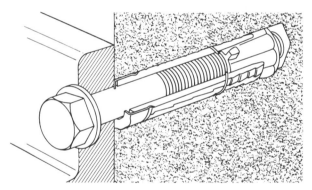

Fig. 3.12   Loose bolt anchor in place

Notes: (a) These devices can be converted to stand-off anchors by using a length of threaded studding in place of the bolt (see Fig. 3.10(a)). The procedure is as follows:

1. Drill a hole of the recommended diameter and depth.
2. Screw the studding into the body of the anchor in accordance with the manufacturer's instructions. At this point a nut and washer should be run on to the studding down to the body of the anchor.

28

3. Push the body of the anchor into the hole.
4. Tighten the nut to set the anchor.
5. Run another nut and washer on to the studding and position in line with the back surface of the subsidiary component (i.e. the stand-off position).
6. Offer up the subsidiary component and secure it with a third nut and washer.

Note: (b) This type of device (and other socket devices) can be set deeper into the base by using another adaptor – a solid collar or sleeve. This takes up the distance between the body of the anchor and the subsidiary component (see Fig. 3.10(b)).

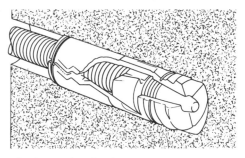

Fig. 3.13   Socket anchor in place

Socket type

1. Drill a hole of the recommended diameter and depth.
2. Screw the studding into the body of the anchor until expansion starts.
3. Push the body of the anchor into the hole using the studding.
4. Tighten the studding to the recommended torque to expand the anchor.
5. Offer up the subsidiary component and secure by running-on a nut and washer.

Note: There is a commonly used alternative device in which the socket is placed in the drilling and then expanded by driving a setting tool into the body of the anchor. This firmly fixes the socket, making it totally independent of any further fixing of the subsidiary component. This is an example in which the expansion anchor is directly comparable with a plug fixing, but with a greater load-carrying potential (Fig. 3.13).

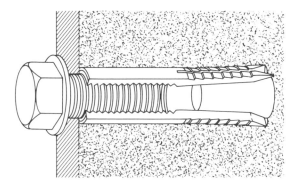

Fig. 3.14 Self-drilling anchor in place

Self-drilling type
This is a version of the socket type of anchor (Fig. 3.14).

1. Fit the anchor in an appropriate adaptor into the chuck of a rotary/impact hammer.
2. Commence drilling.
3. When the chuck almost hits the surface of the base material, withdraw the anchor and clear both anchor and hole of drilling debris and dust.
4. Insert the conical expander plug or wedge into the cutting end of the anchor and hammer the assembly back into the hole without rotating the hammer chuck. It is important only a hammer action is used.
5. Snap off the driving head of the anchor. A hand hammer may be necessary to do this in the case of larger anchors.
6. The anchor is now ready to receive a bolt or stud fixing just as if it were a normal socket anchor.

Expanding anchor (non-friction type)

A heavy-duty expanding anchor which is based on a rather different holding principle is the Leibig Ultra Plus. Instead of the expanding part of the anchor pressing against the sides of the drilled hole and inducing friction between the anchor and the mass walling base – this friction providing the holding power for the anchor – the Ultra Plus expands into an undercut in the concrete, compressing the base upwards against the integral flange of the device set on the surface of the base. No expansion forces are directed into the concrete during setting or the application of load. This gives reliable load-carrying performance, and minimum spacing and edge distances of anchors are possible.

Description The Ultra Plus bolt is made of high-tensile steel with an integral stud (M8, M12 or M16 diameter). At the end of the stud is a round nut which supports the clamping segments. These are opened by a conical sleeve during setting – the segments opening into an undercut in the concrete. Variations in hole depth can be accommodated and embedment depths range from 95 to 190 mm.

Applications Ultra Plus bolts make heavy-duty fixings into a pre-drilled hole on a concrete base. This anchor provides strong, reliable fixings without exerting expansion forces on the concrete. It can withstand shock and dynamic loads, and reduced spacing and edge distances are feasible. Stand-off fixings can also be made using the Ultra Plus (Fig. 3.15).

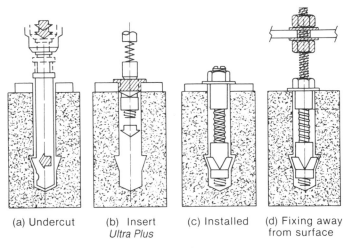

(a) Undercut   (b) Insert   (c) Installed   (d) Fixing away
             Ultra Plus   from surface

Fig. 3.15 Leibig Ultra-Plus anchor setting diagrams

Setting instructions

1. Drill a hole of the recommended diameter and depth. The

hole is to have an undercut rebate in the side of the drilling formed by a carbide undercutter supplied by the manufacturer.

2. Drill a clearance hole through the subsidiary component.
3. Pass the Ultra Plus through the component and into the drilling.
4. Run a nut onto the stud and tighten. This opens the clamping segments and sets the bolt.
5. Apply the specified torque.

Note: Stand-off fixings are placed in the same way, inserting the device in the hole so that its flange rests on the top of the concrete. Tighten the first nut to open the clamping segments and secure the device. Then run-on a second nut with a washer on top and set this at the required stand-off level on the threaded studding. Place the subsidiary component on the washer and fix with another washer and a third nut.

Specialized anchor devices

There are a number of other devices, developed for specialized applications, but which operate on a similar principle to the normal expanding anchors.

Ceiling suspension anchors These have been designed for use below concrete floors or roofs to provide a ring bolt or similar fixing for suspended ceiling systems. They make permanent, non-removable fixings (Fig. 3.16).

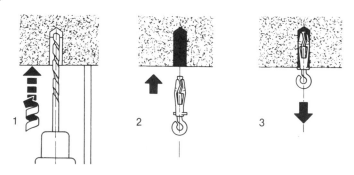

Fig. 3.16 Hilti HA8 R1 suspension anchor setting diagrams

Setting instructions

1. Drill a hole of the recommended diameter and depth.
2. Insert the anchor.
3. Sharply pull down the protruding ringbolt. This expands the anchor and sets it. A claw hammer or screwdriver may be used to pull the ringbolt downwards.

Another proprietary suspension anchor for ceiling duct or pipework fixing is the Fischer L8. Here an 8 mm diameter, 30 mm deep hole is drilled in the concrete soffit, the corrugated section of the anchor is inserted and the sleeve is hammered home to within 1 mm of the anchor plate (Fig. 3.17).

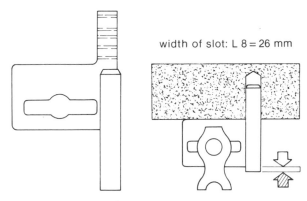

width of slot: L 8 = 26 mm

Fig. 3.17 Fischer L8 ceiling fixing

Insulation fasteners Used to secure non-self-supporting insulation, such as mineral wool, glass wool or expanded polystyrene, to mass walling, these lightweight devices have wide circular or star-shaped heads (Fig. 3.18). Usually they are manufactured from impact-resistant polypropylene and have a finned shank, which grips the sides of the pre-drilled hole in the base when the fixing is tapped home. Its holding power derives from the distortion of the fins.

Cavity wall repair anchors These special devices have been developed to replace corroded cavity wall ties without resorting to pulling down and rebuilding any part of the wall. There are several patterns which are based on the principle of the expanding anchor. Usually they consist of an expanding sleeve placed at both

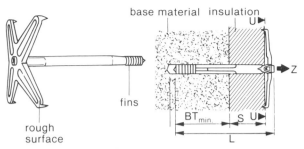

Fig. 3.18 Hilti IN insulation fastener

ends of a steel bar, sufficiently long to span the cavity. Cavity wall repair anchors are usually placed from outside the building, but it is possible to install them from inside if this is more convenient (Figs. 3.19, 3.20).

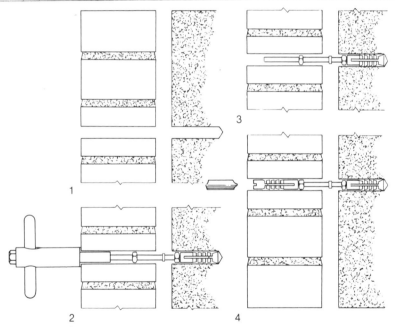

Fig. 3.19 Cavity wall repair anchor (Harris and Edgar);

Setting instructions

1. Drill the recommended diameter hole through the outer leaf of the wall and to a predetermined depth in the inner leaf.

34

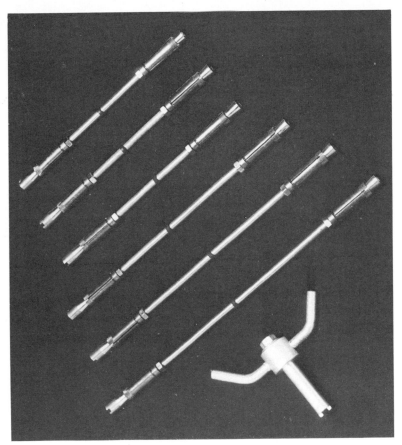

Fig. 3.20 Photograph

2. Insert the steel bar, complete with the inner expansion sleeve, through the outer leaf and to the correct depth in the inner leaf. Hand-tighten using the key provided. This expands the sleeve in the inner leaf. Complete the expansion by applying the correct amount of torque (usually 3 to 7 Nm) to the bolt head with a controlled torque device.

3. Remove the key and insert the outer leaf expanding sleeve over the end of the steel bar and tighten to the required torque using a screwdriver socket attachment, fixed to a torque wrench.

4. Make good the hole in the outer skin.

Note: Where the device does not have a drip collar in the centre of the cavity, the whole device should be set with a slight fall towards the outer leaf of the wall.

A2   Chemical anchors

These fixings are set in a pre-drilled hole, similar to expanding anchors, but they rely on an adhesive to provide their pull-out strength, not the expansion of the sides of the anchor against the face of the drilling.

Chemical anchors consist of two major parts: a resin cartridge (containing the resin and the hardener sealed within separate containers) and the anchor rod (a threaded stud) with its nut and washer. The anchor rod can be replaced by an internally threaded socket if a socket connection, rather than a threaded stud connection, is required.

A chemical anchor can be applied successfully in all locations where an expanding anchor could be used. In addition, because there are no expansion forces involved, it can make strong fixings closer to the edge of the base material than can an expanding anchor.

There is no present evidence to suggest the performance of this type of anchor deteriorates with age.

Further information on the use of anchors in various types of mass walling base, their positioning and other matters affecting their performance, are dealt with in the background section on page 63.

Description Chemical anchors are medium- to heavy-duty fixings which can be applied in dense and strong masonry bases to produce sound structural fixings. They usually consist of a glass phial of resin with a capsule of hardener contained within it. This is placed into the drilling and then a threaded stud, or internally threaded socket, of mild steel or stainless steel is forced into the hole, breaking the phial and mixing the resin and hardener. The fixing exerts no expansion stresses on the base and therefore can also be used for light-duty fixings close to the edge of even low-density concrete blocks.

Anchors are produced by various manufacturers to set studding from 8 to 30 mm diameter and of lengths varying from 110 to 240 mm.

36

Applications Chemical anchors can be used in all locations in which an expanding anchor can be used (i.e. in dense concrete, no-fines concrete, lightweight concrete with a dense texture, stone and brickwork made up of solid bricks without large internal voids). They can also be used in low-density concrete blocks, but high structural performance is only obtained in strong bases. The use of chemical anchors is not recommended in very porous or perforated masonry because the resin tends to become lost in the pores of the base. Fixings can be positioned more close to the edge of the base and generally these anchors can be used in positions subject to vibration or shock. They are unaffected by frost or weather and can be set in damp holes. Some can be installed under water. The drilling of a hole of precisely the correct diameter is essential for the fixing to achieve its full loading capability (Fig. 3.21).

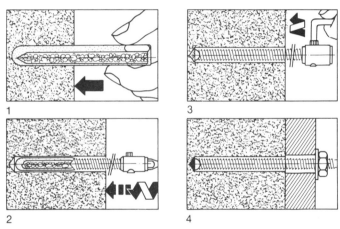

Fig. 3.21 Chemical anchor setting diagrams

Setting instructions

1. Drill a hole of the correct diameter and depth.
2. Insert the resin phial after cleaning the hole of debris and drilling dust.
3. Connect the threaded stud to the chuck of a drill by inserting the hexagonal drive bar in the end of the studding and fitting it into the chuck (in the case of a threaded socket, fit the cap screw and washer provided to the socket and insert the drive bar). Fit the drive bar into the

chuck of the drill. In some types of chemical anchor an adaptor is provided for the drill chuck.

4. Offer the stud (or socket) up to the hole, turn on the machine and force the stud (or socket) through the phial to the bottom of the hole.
5. Stop drilling as soon as the bottom of the hole is reached to prevent over-mixing. (Some studs have a setting depth mark on the shank.)
6. Leave the studding (or socket) in position until the adhesive has set. Then (in the case of the socket) remove the cap screw with a socket wrench.

Note: It is important that the drilling machine has sufficient power to drive the studding to its full depth. Rotary hammer machines are recommended. Do not use the hammer action without rotation.

Only use resin phials which are cool and do not expose phials to direct sunlight.

Specialized chemical anchors

As was the case with expanding anchors, the chemical anchor has been adapted for certain specific tasks.

Cavity wall repair anchors These devices have been developed to replace corroded cavity wall ties without resort to pulling down and rebuilding any part of the wall. There are several patterns available. The inner leaf fixing is usually carried out by chemical anchor, as described above, with an extended threaded stud which bridges the cavity and is gun-grouted into the outer leaf with thixotropic resin. Alternatively, both inner and outer leaf fixings can be made using gun-grouting. It is usual for these devices to be fitted from outside the building, but it is possible to carry out the work from inside if so desired (Fig. 3.22).

Setting instructions

1. Drill the recommended diameter hole through the outer leaf of the wall and to a predetermined depth in the inner leaf. Arrange the drilling so that it passes through a brick and not through a joint.

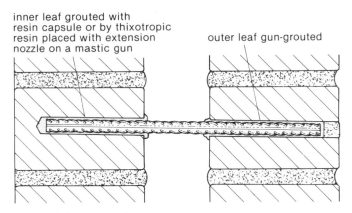

inner leaf grouted with
resin capsule or by thixotropic
resin placed with extension
nozzle on a mastic gun

outer leaf gun-grouted

Fig. 3.22 Chemical cavity wall repair anchor

2. Insert the resin phial into the inner drilling and drive the extended studding into the inner leaf hole using a drill in the same way as the normal chemical anchor. Avoid over-mixing the adhesive.

3. Gun grout the studding in the outer leaf using a thixotropic resin.

4. Make good the hole in the outer brickwork.

Note: When the stud has no drip collar in the centre of the cavity, the whole device should be set with a slight fall towards the outer leaf of the wall. Resin systems can be used with success in damp brickwork, but should not be placed in rain-saturated brickwork.

A3 Injection anchors

The injection anchor is a fixing device which bears a resemblance to the chemical anchors described above. It, however, uses a quick-setting cement which is injected into the drilling through the device itself. It has been developed particularly for use in less dense walling bases.

Description Injection anchors are medium- to light-duty fixings for use in soft mass walling bases. They consist of a hollow metal body through which a quick-setting cement is injected. This seeps out through holes in the body of the anchor to fill an over-large (often dovetailed) drilling in which the anchor is set.

Injection anchors provide a threaded socket to receive bolt fixings from 8 to 12 mm diameter with insertion depths of from 9 to 25 mm; other types are made of nylon to accept woodscrews and coachscrews (4 to 6 mm diameters with anchorage depths of 50 mm). Some anchors, designed for use in bases with large cavities, have retaining nets to prevent loss of cement in the cavities.

Applications Injection anchors can be used in soft mass walling, such as aerated or pumice concrete, hollow brick or blockwork and friable brick or blockwork, to give fixings of predictable strength. They place no expansion forces on the base (Fig. 3.23).

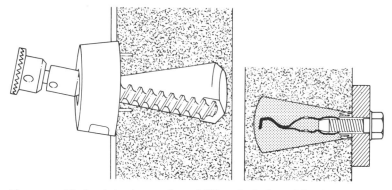

Fig. 3.23   Fischer injection anchor; drilling the hole and finished fixing

Setting instructions

1. Using a special drilling jig, drill a reverse-tapered (dovetailed) hole in the soft base. At first drill horizontally at slow speed until the four centering pins of the drill jig grip the base material. Then enlarge the hole into a reverse taper shape by making circular movements of the drill.
2. Switch off the drill before withdrawing the bit.
3. Insert the anchor until the sealing flange is flush with the wall surface. Insert the protective sleeve into the anchor. This avoids fouling the threaded socket during cement injection.
4. Stir the quick-setting injection cement using a whisk on the drill. The mixing proportions recommended by the

40

manufacturer should be strictly followed.

5. Draw cement into the gun and inject it through the protective sleeve into the anchor body – and through the body into the wall cavities or pores.
6. Allow the cement to set.
7. Pull out the protective sleeve.
8. Fix the subsidiary component with a bolt (or screw).

Note: It is not necessary to drill a dovetailed hole in hollow bricks or blocks, or in friable material.

A4 Screw-in anchors

A fastener that is halfway between a drilled-for and a base-deforming fixing is the screw-in anchor. This requires the drilling of a highly accurate pilot hole, slightly smaller in diameter than the thread diameter of the screw, which is made from hardened carbon steel. This thread-forming screw is then driven into the pilot hole, using the same drilling tool, fitted with an adaptor.

Screw-in anchors have the advantage of making medium- to light-duty fixings without the need for plugs or expensive expanding anchors. The drilling can be performed through the subsidiary component to make a one-operation fixing. The subsidiary component can be removed, if required, after placing.

Description Screw-in anchor screws are made of hardened carbon steel with notched high threads which cut their own groove in the mass walling base. The fixing system comprises the anchor screws, a tool which is fitted to the drill to drive the screw and special highly engineered carbide-tipped drill bits to give consistent and accurate holes. (a most important factor with this fixing). These are supplied free by the manufacturers with purchases of a minimum quantity of screw-in anchors.

Screws have 'Phillips flat countersunk or hexagonal washer heads and range from 25 to 125 mm lengths in two diameters, 5 and 6 mm.

Applications These devices can be used in most solid mass walling bases where quick fixings for metal or timber subsidiary components (maximum thickness 100 mm) are to be made. Both components and base can be drilled in the same operation. Embedment depth should be between 25 and 37 mm. The longer

Fig. 3.24 Buildex Tapcon

Fig. 3.25   One application of BIF Confas

screws in the range are intended for use with the thicker components (Figs 3.24, 3.25).

Setting instructions

1. Drill hole, using the drill bit provided in a hammer drill, through the subsidiary component and into the mass walling base. The depth in the walling should be at least 6 mm deeper than the embedment depth of the screw (embedment depth minimum 25 mm; maximum 37 mm).
2. Snap onto the drill the driving tool with the correct head style and drive the anchor into the hole, while holding the subsidiary component in position. An adjustable depth-sensing nosepiece avoids over-driving the screw anchor. This operates much like a clutch and throws the drill out of gear on contact with the component.

Note: The accuracy of the pre-drilled hole is essential to the good performance of this type of fixing. It is advisable to use the drill bit provided by the manufacturer and replace it regularly as it becomes worn.

Section B: Base-deforming fixings – type 2(b)

Base-deforming fixings in mass walling consist of various types of masonry nail, either driven manually or by means of an explosive or compressed-air tool, and a small group of hammer-in anchors which are used exclusively in aerated concrete bases.

The base-deforming fixing compresses the material of the base into itself as the nail is driven. The compression forces so generated produce friction on the shank of the nail, which resists its withdrawal (Fig. 3.26).

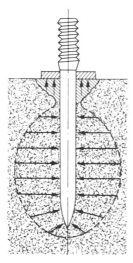

Fig. 3.26 Masonry nail in mass walling; forces generated

For a long time it was considered that, because mass walling bases (unlike timber) were not fibrous, nail fixings into mass walling could never be more than lightweight and relatively unreliable. With the introduction of first the explosive cartridge tool (which we will refer to as the powder-actuated tool) and then the pneumatic tool, the limited performance and application of the masonry nail was immensely increased.

As in the case of drilled-for fixings, there are a number of factors affecting the performance of base-deforming fixings; factors such as the strength of the base material, the way the fixing is driven and the strength of the fixing itself.

B1 Masonry nails

Description These nails require no prior preparation to the base. They are high-quality zinc-plated steel nails which are hand-driven through a subsidiary component (usually wood or light-gauge metal) into relatively soft mass walling bases to make light-duty fastenings. Diameters are usually within the range from 2 to 3.7 mm diameter and up to 100 mm long.

Applications Masonry nails are used to make light-duty fixings into soft brickwork, lightweight concrete with a dense texture and soft stone (Fig. 3.27).

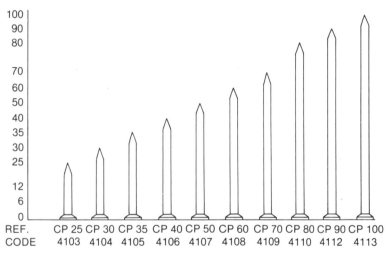

Fig. 3.27 Masonry nails; heavy-duty range by Spit

Setting instructions

1. Drive the nail with short positive hammer blows through the subsidiary component (maximum thickness of a wood component is 50 to 60 mm) to a maximum depth in the base of 20 mm.
2. Nails should be placed between 200 and 300 mm apart.
3. Avoid nailing into mortar joints.
4. Only place one nail in each brick if possible.

Note: It is wise to wear safety glasses while hammering.

45

B2 Powder-actuated fasteners

These fixings require no prior preparation of the base, as do expanding anchors or plugs. There is no pre-drilling and the fixing is merely driven into the base by an explosive charge from a cartridge in the fixing tool or gun. Fixings take two forms (Fig. 3.28):

(a) galvanized steel drive pins for direct firing through the subsidiary component to create a permanent fixing to the base in one operation;

(b) galvanized steel threaded studs for firing into the base to give detachable fixings for the subsidiary component. There are accessories such as ring or eye couplings, clamps and hooks which can be used with threaded studs.

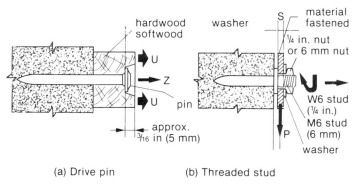

(a) Drive pin   (b) Threaded stud

Fig. 3.28 Fired masonry nail and threaded stud

Both types of fixing achieve their holding power by friction against the shank of the pin.

Typical subsidiary components which can be fired through to make a direct, one-operation fixing to mass walling include:

(a) wood (softwood or hardwood);

(b) steel or aluminium sheeting or brackets;

(c) soft sheet materials (insulation board, wood wool, cork, etc.) usually employing an additional washer to avoid punching through.

Different sizes and shapes of drive nail are produced by the manufacturers to match the character and thickness of the subsidiary component material.

Further information on the use of powder-actuated fixings in various types of mass walling base, their positioning and other matters affecting their performance, are dealt with in the background section on page 63.

Description Powder-actuated fixings are drive pins or nails of hardened (austempered) steel fired from a tool by an explosive cartridge to make a permanent fixing into a mass walling base. Pins are produced with various head and shank styles to suit the material with which they are used. Shank diameters are 3.5 or 4 mm and lengths are up to 100 mm. Plain headed pins are intended to fix the subsidiary component direct to the base. Threaded stud-headed pins are in diameters from 6 to 10 mm. These are intended to provide a firm anchorage to make removable connection points for the subsidiary component. There are various colour-coded strengths of cartridge available to give the necessary force to drive the pin into various bases.

Applications Powder-actuated pins can make medium- to heavy-duty fixings in dense concrete, lightweight concrete with a dense texture, brickwork and stone, such as sandstone, soft dolomite and shaly bedded rock. Fired fixings are not suitable for dense concrete of extreme strength (over 60 N/mm^2), no-fines concrete, aerated concrete, perforated or porous clay or calcium silicate bricks (below 10 N/mm^2), hollow clay blocks or hard calcite stone and granite. Care needs to be taken when using these fixings with hollow bricks, calcium silicate bricks and hollow lightweight concrete blocks. In these cases preliminary trial fixings should be made if the performance of the fixing is critical.

The success of these fixings depends on the type of fixing tool, positioning of the pins and the thickness and type of the subsidiary component.

Fixing tools

Because the drive pins are considerably harder than the base material into which they are driven, the powder-actuated tool has to ensure that the pin is aligned and guided throughout driving to prevent sideways deflection.

There are two types of driving tool: one direct-acting, the other indirect- (or piston-) acting.

The direct-acting tool gives high velocities with entry speeds up to 500 m/sec and the pin is driven by the expansion of gas as a result of the firing of the cartridge. The fastener propulsion and depth of penetration is not controlled.

The indirect-acting tool operates by released gas acting on a piston in the barrel of the tool. This drives the pin into the base under controlled conditions at velocities up to a maximum of 100 m/sec. The piston is retained within the tool, being stopped at an appropriate point to give the required depth of penetration.

Both types of tool have safety devices which do not allow firing except when the barrel is in firm contact with the base. If the tool is 6° or more out of perpendicular to the base, one type of tool will not fire.

The depth of penetration into the base should be within 22 and 32 mm and this can be achieved by adjusting the strength of the cartridge until the penetration is correct.

Powder-actuated tools are sophisticated pieces of equipment and should be used with respect. Tool, pin and cartridge should all be elements of the same fixing system, their quality and sizing being designed to guarantee safety for the operator and successful fixing performance. There is no room for operator improvisation.

Some manufacturers offer training courses in the use of their equipment and the Powder Actuated Systems Association produces basic safety information. Operating instructions should be followed meticulously: only fasteners and cartridges compatible with the tool should be used and only skilled operators should have charge of the equipment.

Positioning of pins

The minimum distance between the pin and the edge of the component should be 50 mm. The minimum spacing of pins should be two times the penetration depth of the pin.

Thickness and type of subsidiary component

When the component is fixed direct to the mass walling, its thickness and type is critical to the success of the fixing. If the material is too thin, the pin may punch through (as in the case of thin metal) or splinter (as in the case of wood). If the material is too thick, the fastener could bend, or not penetrate the base sufficiently, even using a pin with the longest shank.

Manufacturer's data sheets give full details of the thicknesses of subsidiary components which can be fixed using various lengths of fastener.

When soft materials are being fixed, such as insulation board or thin metal sheets, with the risk of pull-over failure or firing through, additional washers may be required.

Setting instructions
Always follow the manufacturer's operating instructions minutely.

1. Select the length and type of pin required from the manufacturer's data, bearing in mind the material of the base and the type and thickness of the subsidiary component.
2. To ensure the correct cartridge is used to give correct penetration depth, commence with a test firing, using a low-strength cartridge. If the pin does not penetrate the base sufficiently (22 to 32 mm) change to a higher-strength cartridge. The various strengths of cartridge are colour coded. (As a general guide, the harder the concrete, the less penetration of pin is required; the softer the concrete, the greater the penetration and the longer the shank length required.)
3. Place the tool firmly against the base (or, if firing through the subsidiary component, the component), ensure the tool is at right angles to the base, and fire.
4. When firing into concrete or stone, the use of a spall stop is advisable (Fig. 3.29). This is a steel ring on the end of

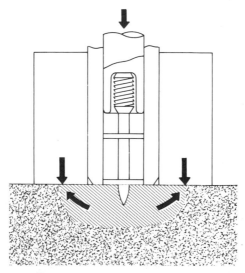

Fig. 3.29 Spall stop

the tool which exerts a supporting downward pressure on the surface of the base during firing. This encourages the compression of the base (and improves holding power) and reduces the danger of spalling.

Note: It is wise to wear safety glasses while making powder-actuated fixings and, in enclosed surroundings, ear protection too. When working on thin partitions, it is recommended that the space on the other side of the partition is fenced off in case of firing through.

Further safety advice is contained in the Health and Safety Executive guidance note PM14 *Safety in the Use of Cartridge Operated Fixing Tools*, the Powder Actuated Systems Association's *Guide to Basic Training* and in *Site Safety* by J. C. Laney in this Site Practice series.

B3 Pneumatic-actuated fixings

These fixings are identical to powder-actuated fasteners; the only difference is in the method of driving the pin. All parts of the preceding section can therefore be considered as applicable to pneumatic-actuated fixings, with the exception of that part dealing with the fixing tools.

In the case of pneumatic-actuated fixings, the pin is propelled by a piston driven by the release of compressed air through the tool. This gives a low-velocity, controlled impetus to the pin, unlike the direct-acting powder-actuated tool. Clearly the disadvantage of a pneumatic fixing tool is that it needs a compressor on site with a long air line connected to the tool. If this facility is available, however, the speed of placing pins can be greater than with a powder-actuated tool. The pins in the pneumatic tool are collated into strips of ten and fed into the magazine two strips at a time.

For the larger project, the pneumatic tool has productivity advantages. To these can be added other benefits, such as less recoil and less noise on firing. The air pressure can easily be regulated through a range from 6.2 to 10.5 bar to give minute adjustment to the penetration of the pins – a less clumsy process than the test firings with a variety of different-strength cartridges often needed to achieve the correct penetration of powder-actuated pins.

Although both types of tool have safety devices to prevent accidental firing, there is a feeling that the pneumatic tool is safer to use largely because of the absence of explosive cartridges. It does, however, tend to have mobility problems due to the need for an air line. This has led to the use of compressed air cylinders for smaller fixing jobs. Pressure cylinders, similar to those used in the aqualung, can be carried on the operative's back allowing approximately thirty fixings per cylinder to be made.

Note: The wearing of safety glasses when using pneumatic-actuated tools is advisable.

B4 Hammer-in anchors

This small group of mass walling fixings has been introduced to make speedy and relatively light-duty fixings to low-density walling materials without the need for drilling.

Description Hammer-in anchors are made of zinc-plated steel with standard, looped, countersunk screw or hexagonal nut styles of head. Diameters of 5 and 6.5 mm are available and lengths from 50 to 150 mm.

Applications These anchors are designed specifically for making quick, low-cost and relatively light-duty fixings in aerated concrete without the need for drilling the base. The holding power of the fixing will be affected by the composition, density and compressive strength of the aerated concrete. Fixings can be made straight through softwood subsidiary components up to 75 mm thick, otherwise the component should be drilled (Fig. 3.30).

Setting instructions

1. Pass the anchor through the subsidiary component (if pre-drilled).
2. Insert the pin of the setting gauge into the hole in the head of the anchor.
3. Hammer the head of the gauge until the stop of the gauge comes into contact with the component.
4. Remove the gauge and drive the anchor the rest of the way home with direct blows of the hammer. This action will splay out the anchor sleeve over the expander in the base.
5. In congested situations, where positive hammer blows are difficult to deliver, a special setting tool, supplied by the manufacturer, can be used.

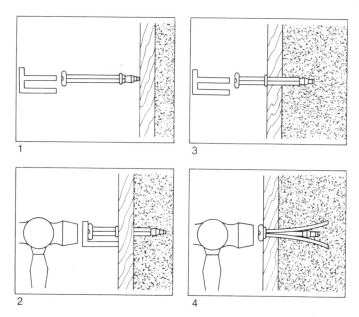

Fig. 3.30 Loden hammer-in anchor setting diagrams

Note: It is possible to remove standard and looped-head types of anchor by inserting a long, loose-fitting nail (supplied by the manufacturer) into the hole in the head of the anchor and driving the expander, within the anchor, free of the anchor sleeve. The sleeve can then be withdrawn, using pliers. Some damage may occur to the component and the surface of the base material.

Section C: In situ fixings – type 2(c)

Unlike the two preceding groups of fixings (types 2(a), 2(b)) in situ, devices are placed at the time of constructing the mass walling base. Their performance depends, as did the previous categories, on the bond between the fixing and the base, as well as the character of the base itself. In addition, because the in situ fixing is placed at the time of constructing the mass walling base, its success also depends on the accuracy of its positioning in the base and its maintenance in that position until the base materials have hardened. This requires considerable foresight in the construction process and care to protect the fixing during and after construction.

C1 Built-in fixings

Built-in fixings include such items as joist hangers and anchorage straps, designed to connect a subsidiary component to a mass walling base, and those used, like cavity wall ties, cramps and dowels, to connect individual parts of a mass walling base together. The former type replaces the older practice of building in subsidiary components, or ledging them on a corbel formed in the walling. The latter are present-day examples of a long line of historic lead cramps and iron dowels in masonry. All these achieve their holding power by adhesion between the fixings and the mortar in which they are bedded and the weight of the walling above the fixing.

Because of the need to bed these devices, their positioning is limited by the coursing discipline of the walling.

In the case of some fixings of this type, manufacturers produce tables of safe working loads for the device when used in specific types of walling material.

Further information on the use of in situ devices in various types of mass walling base and matters affecting their performance are dealt with in the background section on page 63.

Joist hangers

Description These are saddles, usually made of 2.7 mm thick galvanized steel with a build-in flange (sometimes fish-tailed). They are shaped to form a seating for either one timber floor or roof joist, or two similar joists on opposite sides of a partition wall (Fig. 3.31). Various sizes of hanger are available to suit joist sizes, from 100 to 250 mm deep and from 38 to 150 mm wide.

Applications Joist hangers are used to make connections of predictable engineering performance between timber joists or beams (installed after the building of the wall) and walling made up of coursed elements, such as bricks or blocks. These should not be confused with joist hangers without the build-in flange and designed to be connected to the base by other anchorage devices, such as expanding anchors. Such hangers are accessories to the main fixing device (see section D, p. 61). Also there is a range of similar hangers used to make timber-to-timber connections (see Ch. 4).

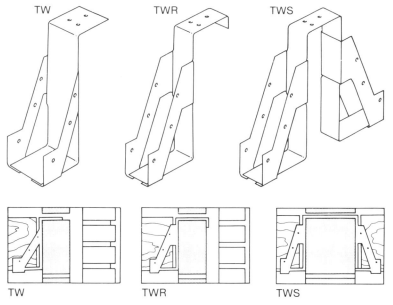

Fig. 3.31 Three versions of the Catnic TW joist hanger

Setting instructions

1. Joist hangers should be accurately positioned along the appropriate walling course, their flanges being placed on top of the course. Ensure the flanges are adequately surrounded by the mortar of the bedding joint.
2. Do not load the hangers (particularly single hangers) until their flanges are secured by higher lifts of walling and the mortar of the joint has completely hardened.
3. Later set the joist in the saddle and fix to the hanger with 30 mm × 375 mm square twisted sherardized nails through the pre-punched holes in the hanger.

Anchorage straps

Description There is a variety of patterns for these straps which are designed to withstand the effects of wind pressure on the structure and perform general bracing duties. They are mostly manufactured of 5 mm galvanized mild steel for horizontal restraint applications and 2.5 mm for vertical restraint. They are pre-punched (usually at 12.5 mm centres) to receive nail fixings to the subsidiary timber components and there is a variety of shaped ends for building into the walling (Fig. 3.32).

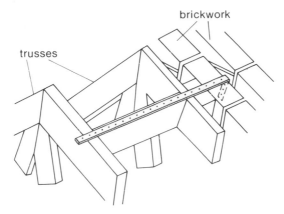

Fig. 3.32   Bev strap bracing roof trusses

Applications Anchorage straps are used to provide stiffening and lateral restraint to roof or floor joists running at right angles or parallel to the wall (horizontal restraint) or holding down wall plates and trusses (vertical restraint). Lateral restraint straps are sometimes used in combination with joist hangers.

Some vertical restraint straps are not built into the wall but are connected to it by masonry nails or screws into plugs. These straps are accessories to the fixing device (see section D, p. 61).

Setting instructions

1. Straps should be accurately set out as required by the building design, ensuring that their flanges are adequately surrounded by the mortar of the bedding joint.
2. Only attach the strap to the timber members when the mortar has had sufficient time to harden.
3. Nail connections to the joists should be made using 30 mm × 375 mm square twist sherardized nails.

Cavity wall ties (and other wall ties)

Description Galvanized steel, stainless steel or plastic wall ties are produced in a variety of patterns and sizes for tying the two leaves of a cavity wall together, or tying a thin stone or precast concrete facing to a backing wall. Cavity wall ties are most commonly produced in two forms: the galvanized wire butterfly tie, and the twisted fish-tail flat tie made from galvanized steel strip. Today these have been joined by polypropylene ties and ties with retention devices to hold insulating batts within the cavity against the inside leaf (Fig. 3.33).

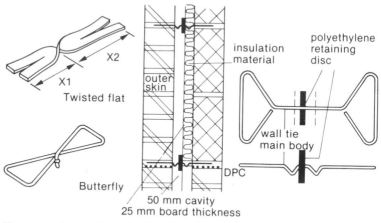

Fig. 3.33 Butterfly, twisted and insulation-retaining wall ties

It is worth noting that the condition of these ties can never be easily monitored, hence corrosion could occur and remain undetected for a long time. The thickness of galvanizing required by BS1243 has recently been increased due to the extensive corrosion suffered by wall ties galvanized in accordance with earlier requirements.

Applications Wall ties are used to tie parts of a mass walling base together, such as the two leaves of cavity walls, or thin claddings to brick or block backing walls. In the former case the ties are spaced at 900 mm centres horizontally and 450 mm centres vertically in a staggered pattern with greater frequency at wall openings; in the latter case ties will be positioned according to the pattern of the cladding joints and will have a dowelled or lipped end for fitting into holes or slots in the edge of the cladding slabs (see Fig. 3.34)

Setting instructions

1. Set the wall ties in the mortar bedding joint as the wall is

56

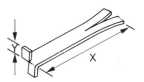

Fig. 3.34 Lipped tie

built, ensuring that each tie does not fall towards the inner leaf. An outward fall is preferable.

2. In the case of insulation retaining ties; build the inner leaf 450 mm higher than the first row of wall ties; set the insulation batt resting on the ties and between the retainer and the inner leaf; build up the outer leaf and set the next row of wall ties.

Cramps and dowels

Description A variety of galvanized and stainless steel cramps and dowels is available of many sizes and shapes. Cramps are usually made from steel flats, often bent to fit into grooves in the masonry elements; dowels are usually made from steel rod (Fig. 3.35).

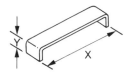

Fig. 3.35 Coping cramp

Applications These are used for reinforcing the joint between two adjoining stones or precast concrete elements, such as coping stones.

Setting instructions

1. Bed cramp (dowel) in slot or mortise cut or cast in the stone or concrete element.
2. Point with mortar.

C2 Cast-in fixings

Many of these anchorage devices (such as plugs and sockets) are similar to the drilled-for fixings (section A1 of this chapter) in concrete bases, except that cast-in fixings are placed at the time of pouring the concrete. Their precise position, therefore, has to be established early in the construction process. They have to be positioned either by wiring them firmly to the reinforcement, or by nailing them to the inside of the formwork with light, outward-pointing nails which will pull off when the formwork is struck, leaving the fixing device in the face of the concrete.

An alternative method of installation, and one which allows for minor positional adjustment at a later stage, is to form a pocket or mortise in the concrete by casting in a polystyrene or other soft material block. After the concrete has hardened, the block can be picked out leaving a recess in the concrete. The fixing can then be accurately positioned in the mortice and bedded in place using concrete or one of the many proprietary concrete grouting compounds. The alternative method allows for greater accuracy of positioning and avoids the fixing being displaced during concrete pouring; but care is needed to ensure that all traces of the temporary block are removed from the pocket and that the grouting bonds properly to the surrounding concrete.

Plugs and sockets

Description Traditionally, cast-in plugs were of wood – a small timber block lightly nailed to the inside of the formwork, so that when the formwork was struck, the block would remain set in the concrete. Today these lightweight fixings are usually made of rotproof plastic. Unlike timber, they will not shrink or swell and will receive a nail or screw fixing without the danger of splitting. They also protect the nail or screw from corrosion. Depths of proprietary plugs vary from 25 to 38 mm and they are usually wedge-shaped to afford greater pull-out resistance.

More heavyweight fixings are made by casting in a ferrous or non-ferrous socket to receive a bolt (size from M10 to M24) (Fig. 3.36).

Applications Cast-in plugs give light-duty fixing points for nails or screws, while cast-in sockets attend to the need for heavier fixings in precast or in situ concrete. They receive a bolt fixing for the subsidiary component or accessory.

Setting instructions
These will vary with the type of plug or socket used and how it is

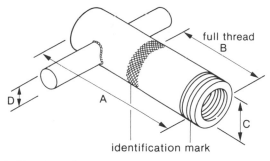

full thread
B

D

A

C

identification mark

Fig. 3.36 Cast-in socket

to be secured within the base before pouring the concrete; i.e. whether wired to the reinforcement, etc.

Cast-in channels

Description There is a number of proprietary cast-in channel systems available, which are profiled to receive a variety of T-headed bolts or ties. The position of these accessories is clearly adjustable in one direction at least – parallel to the length of the channel. Channels and accessories are made of mild steel or stainless steel and in a wide variety of sizes and loading capacities. They are supplied with polystyrene filling inside the channel to avoid fouling during the pouring of concrete, and have several patterns of rear anchorage straps to improve their holding power (Fig. 3.37).

Applications Cast-in channels provide reliable restraint and loadbearing anchorage for a variety of matching accessories, such as T-headed bolts and ties, which are used to connect thin cladding panels direct to the concrete base, or loadbearing accessories, such as corbels.

Setting instructions
Once more, detailed procedure will vary with the particular system used.

1. Carefully mark out the position of the channel on the back of the formwork.
2. Depending on the type of channel and the material of the formwork, lightly nail or staple (supplied by the

59

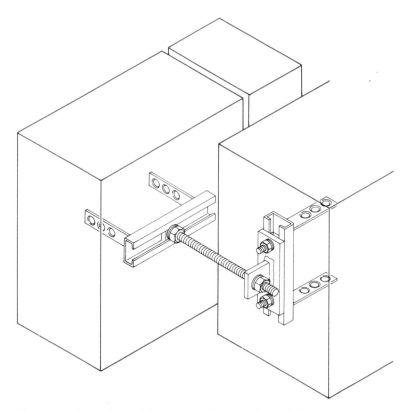

Fig. 3.37 One of the Halfen range of cast-in channel fixings

manufacturer) or bolt the channel in position.

3. Before the formwork is struck, unscrew the bolt nut.
4. Clean out the polystyrene filling from the channel. It is now ready to receive its matching accessory.

Cast-in corbels

Description Manufactured in a variety of ferrous and non-ferrous metals, these heavy-duty fixings come in a variety of shapes – flat, cranked, flat fish-tailed, cranked fish-tailed, or hoop-shaped (frame corbel) (Fig. 3.38).

Applications Corbels are loadbearing fixings used primarily to support heavyweight cladding panels.

60

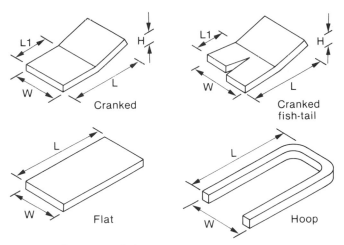

Fig. 3.38 Types of cast-in corbel

Setting instructions
Cast-in corbels are usually grouted into pre-formed pockets in the concrete, using either a concrete grout or a proprietary grouting compound.

Section D: Fixing accessories

A wide variety of accessories is used in combination with mass walling anchorage devices, such as expanding anchors or cast-in sockets. These accessories form an intermediate fixing between the device, which makes the anchorage in the base, and the subsidiary component.

An example of this is a range of loadbearing angle corbels or continuous angles used to carry heavyweight claddings (Fig. 3.39). These depend for their performance on anchorage supplied by expanding or chemical anchors, cast-in sockets, cast-in channels or similar fixings. A joist hanger without a build-in flange is a similar accessory, depending as it does on the anchorage provided by screws and plugs or masonry nails (Fig. 3.40). In addition, many anchorage straps are not built into the brickwork or blockwork, but rely on masonry nailing or similar fixings. These, too, are accessories.

There is a number of specialist accessories that concern the tying back of brick facing to concrete structure. Some, like the Harris

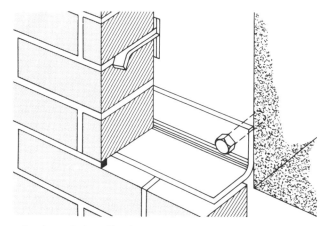

Fig. 3.39 Angle corbel application

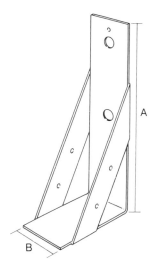

Fig. 3.40 Joist hanger without build-in flange

and Edgar sliding brick anchor, are designed to allow the differential movements of brick and concrete to take place without imposing stress on either material (Fig. 3.41), but the fixing is connected to the structure by expanding anchors, making the device an accessory. Similar accessories include bolted fish-tail ties and dowel ties, a few examples of which are illustrated (Fig. 3.42).

Also within this category are the matching accessories for use with cast-in channels and the innumerable pipe clamps and pipe

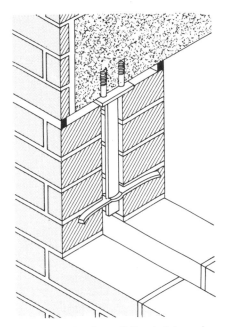

Fig. 3.41 Harris and Edgar sliding brick anchor

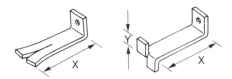

Fig. 3.42 Bolted fish-tail ties etc.

support systems, which once more form a bridge between the fixing device and the subsidiary component.

Section E: Background to fixing performance in mass walling

There is a number of factors which affect most medium- and heavy-duty fixings in mass walling bases. These have been referred to briefly and in a general way in the preceding sections of this chapter. For those who require more detailed background information, this section has been included.

Fixings in dense concrete bases

The following factors will influence the fastener's performance in dense concrete bases:

(a) compressive strength of the concrete;
(b) size of the concrete's aggregate;
(c) pH value of the concrete;
(d) reinforcement pattern in the concrete;
(e) fastener spacing in relation to base dimensions.

Compressive strength of the concrete

The higher the compressive strength of the concrete base, the greater the holding power achieved by all types of anchor fixings. The same is true of fired fixings. In their case, however, there are other factors involved. The stronger the concrete, the heavier the power load required to drive the fastener and the higher the failure rate due to spalling. Usually the compressive strength of structural concrete is around 25 to 40 N/mm^2. This range is suitable for both types of fixing. A fired fixing, though, should not be used in extremely strong concrete (higher than 60 N/mm^2). It is interesting to note that in the case of the chemical anchor, if the concrete strength is below 35 N/mm^2, a core of concrete is likely to break away with the chemical anchor, rather than the adhesive fail. Above that strength, failure will probably occur in the anchor itself rather than in the adhesive.

It is particularly important that the strength of the concrete is established where heavyweight anchors are being used which are expected to produce reliable structural performance. Manufacturer's performance data are compiled for fixings in stated grades of concrete. If the concrete involved is of a lower grade, the expected performance of the fixing must be reduced accordingly. It may even be necessary to carry out tests to establish the true strength of the fixing in a particular concrete.

Always bear in mind that the deeper the anchor, the greater the masonry support it receives and the greater its pull-out strength. The greater the depth of penetration of a fired fixing on the other hand, the greater the failure rate and the danger of spalling. The optimum depth of fired fixings is 22 to 32 mm.

Also it should be remembered not to set drilled-for anchors in (or fired fixings into) fresh concrete less than 7 days old. If anchors are set in fresh concrete 7 days old or slightly older, they should not be fully loaded until the concrete has been allowed to reach its full compressive strength – usually 28 days for normal Portland cement concrete.

Size of aggregate

As the size of aggregate increases, so the minimum spacing and edge distance of expanding anchors must be increased to avoid spalling and failure in holding power. The guide dimensions given under 'Fastener spacing in relation to base dimensions' below are based on concrete with average-sized aggregate for a structural concrete.

If a fired pin strikes a large, hard piece of aggregate during driving, it may be deflected or bent. There is also a risk of spalling. The use of a spall stop – a metal disc on the nose of the gun or tool – is advisable.

pH value of the concrete

Concrete with a pH value below 7 or above 11 is extremely corrosive to zinc-plated steel anchors and masonry nails or pins. Normal Portland cement concrete without additives has a pH value between 9 and 11 and presents no corrosive hazard, but the presence of some additives can alter this situation.

Concrete in the neutral range between 7 and 9 can, under certain atmospheric and climatic conditions, be corrosive.

Any zinc-plated fixing set in a corrosive concrete should be considered as having a limited life. However, there are stainless steel and aluminium-bronze fixings available for use in corrosive conditions originating either in the base or from aggressive environments such as chemical plants, sewage works, etc.

Reinforcement pattern in the concrete

No drilling of holes, placing of anchors or firing of pins should be carried out near the reinforcing wires of prestressed structural concrete. In other types of concrete, expanding anchors should, as far as possible, be placed between the reinforcing bars. In heavily reinforced concrete it is possible to reduce the minimum spacing of anchors (see p. 66), but in coarse (large aggregate) concrete it would be increased.

Fired fixings in normal reinforced concrete do not generally impair its strength. However, fasteners should always be placed so as to avoid striking reinforcement, particularly in narrow structural members where the cross-section is likely to be heavily reinforced.

Fastener spacing in relation to base dimensions

Both the properties of the concrete and the anchor design govern the minimum anchor spacing, the minimum distance from the edge of the base to the anchor and the thickness of the base into which the anchor can be set.

Anchors which do not expand mechanically and exert a force on the base material (cast-in sockets, plugs and lightweight fixings generally) can be placed at closer minimum spacing than is recommended here, assuming their load is not large. Expanding anchors should never be placed too close to the edge of the base, otherwise cracking, spalling or breakaway could occur. It is interesting to note that the performance of two expanding anchors placed too close together can be dramatically reduced.

To achieve the best reliable fixing in average concrete using expanding anchors the following spacings are recommended by one manufacturer based on the length of embedment of the anchor (the dimensions in brackets represent the minimum spacing recommended with *reduced* performance):

1. Distance centre-to-centre of anchors: 2 to 3.5 times anchor length (1 to 2 times anchor length).
2. Distance between centre of fastener and edge of base:
   (i) with load at right angles to edge: 2.5 to 4 times anchor length (1 to 3 times anchor length);
   (ii) with load parallel to edge: 1.5 to 3 times anchor length (1 to 3 times anchor length).
3. Thickness of concrete base in relation to anchor length: 2 times anchor length.

In the case of chemical anchors, because there is no expansion involved in their placing, distance 1 can be reduced to 1 times the anchor length (with 0.5 times the anchor length as the minimum with reduced performance) and both distance 2(i) and 2(ii) are 1.5 times the anchor length (0.5 times the anchor length as the minimum with reduced performance).

Other manufacturers give alternative advice based on the diameter of the fixing. The range of dimensions given above results from the different types of expanding anchor available. It is, therefore, always advisable to follow the recommendations of the particular manufacturer concerning the actual device being used.

In the case of fired fasteners, the distance between the pin and the edge of the concrete should be 50 to 100 mm (depending on the grade of concrete and the shape of the base); the centre-to-centre distance of pins should be 2 times the depth of penetration of the pins; and the depth of penetration of the pin should never exceed half the thickness of the base.

There are two further influences on the performance of chemical anchors – the effects of temperature and chemicals.

Effect of temperature

If chemical anchors are exposed to consistently high temperatures (in excess of 60 °C) they will show a decline in holding power. This, however, is a rare environment and one that should not often be encountered. Temperature, too, will affect curing times. Polyester-based resins can vary in hardening time from 10 minutes at 20 °C and over, to 1 hour at 0 to 10 °C, or 5 hours at minus 5 °C.

Influence of chemicals

Once more this will vary depending on the resin used in the device. Generally chemical anchors are resistant to most common acids and alkalis in normal concentrations. If there is a particular hazard in the location of the fixing, it would be prudent to check with the manufacturer.

Fixings in lightweight concrete bases

Generally, as far as lightweight concrete with a dense structure is concerned (i.e. concrete with lightweight aggregate as opposed to aerated or foam concrete), the considerations applying to dense concrete hold good. The proviso is, of course, that because the concrete has a lower compressive strength (up to 30 N/mm^2), the fixings will have proportionately reduced performance. In concrete of 1800 to 2000 kg/m^3 density, it is not advisable to use anchors with a bolt diameter over 16 mm. It is also unwise to use self-drilling anchors in any lightweight concrete.

Fired fixings make only the lightest duty fixing in concrete of a compressive strength less than 10 N/mm^2 and should not be used at all in aerated concrete.

Aerated concrete does produce an additional problem. Because of its large voids, it tends to assist the corrosive attack of some of its additives on zinc-plated anchors. As the strength of this type of concrete is only up to 8 N/mm^2, only lightweight anchorage fixings are usually considered, and therefore plastic plugs (such as the Hilti H6 and FD anchor, or the Rawlnut) are usually adequate.

Expanding anchors are not recommended for use in aerated concrete.

Fixings in brick and clay block bases

Because bricks and clay blocks are usually brittle, only fixings exerting relatively low expansion forces should be used. Fired fixings are not suitable for use in perforated bricks or hollow blocks.

The following factors influence the fastener's performance:

(a) brittleness and character of the bricks or blocks;
(b) position and size of the cavities in the walling units;
(c) character of the walling.

Brittleness and character of the bricks or blocks

If the brick is extremely hard and brittle, anchors can be set in the mortar joints, but widely varying pull-out performance may result, depending on the quality of the mortar. Tests should be carried out to determine the achievable holding power of such a fixing.

Otherwise anchors should be placed in the centre of the brick or block; but even then, if the fixing is expected to carry a heavy load, loading tests are advisable.

Position and size of cavities in the base

Cavities can considerably reduce the bond surface between anchor and base. They, therefore, must reduce the holding power of the fixing. If possible, the anchor should pass through as many walls between internal cavities as is practicable in order to achieve the best pull-out performance. Nevertheless, however good the performance is, it will never match that of a similar fixing in a solid brick or block. Injection anchors (section A3 of this chapter) have been developed specially to overcome the difficulties of fixing to this type of base. Fixings to some hollow clay blocks tend to fall more naturally within the scope of Chapter 6, 'Mechanical fixings in cellular bases'.

Character of the walling

The width and strength of the mortar joints affect the strength of the wall, as well as the holding power of any anchorage fixing placed in it. If the joint is of hard mortar, good anchoring is achieved even if the brick cracks during expansion of the anchor. If, on the other hand, the joint is weak, the performance of the anchor is lost if the brick cracks.

It is unwise to use any self-drilling anchors in brick or clay block bases.

4

Mechanical fixings in timber bases

Most fixings in timber bases fall into two of the fixing types listed in Chapter 2 – base-deforming fixings – type 2(b) – and through-fixings – type 2(a) (p. 92). The base-deforming fixings include woodscrews, in spite of the fact that they need a pre-drilled pilot hole. This is because a screw's primary holding power is achieved by its thread cutting a matching groove in the timber, i.e. deforming the base.

Section A: Base-deforming fixings – type 2(b)

Because of timber's relatively soft, non-brittle and fibrous nature, most fixings which have evolved for use with timber bases are those which are driven or otherwise forced into the timber, deforming its fibres in the process. It is the grip which develops between the timber fibres and the shank of the fixing which gives the device its holding power and allows it to make an anchorage in the wood.

This group of timber fixings includes nails, woodscrews and toothed metal plates.

The pull-out strength of such fixings as plain shanked nails is considerably affected by the type of wood and its moisture content; while the performance of other fixings, such as annular ring shanked nails and screws, is less influenced by the moisture content of the wood. Screws generally produce a greater withdrawal resistance than nails; and annular nails than plain shanked nails.

A1 Nails (including staples)

These fixings generally require no prior preparation work to the subsidiary component or the base. Some thin timber components, however, may have to be pre-drilled to avoid them being split when the nail is driven; and metal or brittle components (such as roof tiles) have to be punched or fabricated with pre-formed holes to receive the nails. It is sometimes said to be advisable to pre-drill pilot holes up to 80 per cent of the nail shank diameter when nailing some hardwoods (afrormosia, ash, greenheart, beech, gurjun/keruing, iroko, opepe, sapele, teak, jarrah and karri) but strong structural connections can be made with these timbers using plywood gussets without drilling holes.

Nails can be driven by hand or by pneumatic nailing machine. These machines are becoming increasingly common on site where compressors are in use. They either drive nails specially manufactured for machine nailing and collated into strips or coils, or similarly arranged batches of staples. Nails usually have a round, semiround or tee head. Staples can either be equal- or unequal-legged. The air pressure can be adjusted to countersink the nail, or not, as required. There are smaller manually-operated staple machines for lighter stapling jobs and a mallet-actuated machine for nailing floor boards. Nailing machines clearly increase productivity. Up

Fig. 4.1 Coil nailers with magazine capacities of up to 400 nails from BeA

to 180 nails per minute can be placed using some pneumatic machines (Fig. 4.1).

The nail is one of man's oldest forms of mechanical fixing device, being a metal equivalent of the medieval carpenter's peg or dowel. Once nails were forged and then they were cut from metal plate (cut nails). Eventually, however, with the introduction of wire into nailmaking, the industry was revolutionized. Today, most nails are made from mild steel wire (wire nails), although copper and aluminium alloy wires are also used. There are still a few surviving cut nails, made from black rolled steel to produce square-sectioned nails, such as floor brads.

There is a considerable number of different types, sizes and head styles in nails, largely developed for the particular type of subsidiary component to be fixed. More detail of the structural use of nails will be contained in the background section of this chapter on page 104.

Description Mostly nails are driven through the subsidiary component and into the base in one operation. There are a few exceptions, such as the glazing sprig. Here the nail does not pass through the subsidiary component, but is driven into the base alongside the component, trapping it against a projection in the base.

Nails for use in timber bases come in a wide range of sizes and types, capable of making all strengths of joint from light- to heavy-duty. The major types of wood nail are set out and described in Table 4.1. The shape of head and shank are largely determined by the type of subsidiary component to be fixed; the size and shape of the shank is primarily dictated by the structural performance required. While most nails are made from mild steel or plate, others are made from non-ferrous metals (copper or aluminium alloys) for use where corrosion could be a problem. Also some nails, such as escutcheon pins, are manufactured in non-ferrous metals because they are intended to be seen with decorative ironmongery.

It will be observed that nail diameters are given in millimetres. This practice has now largely superseded the habit of quoting nail diameters in standard wire gauge, although s.w.g. sizes may still be encountered from time to time. A table of equivalent diameters is given in Appendix 3.

Applications Nails are used to make joints of varying

structural performance between pieces of wood, or between a variety of subsidiary components (such as roof tiles, bituminous felt, plasterboard, metal fixing straps and light metal accessories) and a timber base. The selection of the right nail for the job will be influenced by: (a) the type of subsidiary component; (b) the performance required from the fixing; (c) the appearance required; (d) the likelihood of corrosion. Brief advice on the correct use of various nails is given in Table 4.1.

Type of subsidiary component

The avoidance of pull-over failure, in which the subsidiary component is torn over the nail head, often dictates the size and style of the nail head; for instance, soft, thin or flexible sheet materials, such as roofing felt, need to be fixed with large, flat circular headed nails. In the same way, when a nail is driven through a thin timber component, there is a danger that it will be split, unless the shape of the shank of the nail is selected carefully. An oval-shanked, lost-head nail could be used in this case, placed so that the oval shank aligns with the wood grain of the component. Specialist nails like corrugated sheeting nails may have round-dome heads with washers to discourage rainwater leakage round the nail shank.

Performance required of the fixing

The engineering performance of the nail fixing will clearly be influenced by the nail type and size. Generally, round-headed plain wire nails are used for carpentry, timber-to-timber joints. Thicker-gauge nails withstand lateral loads better than thinner nails. The longer the shank of the nail (within reason), the greater is its pull-out resistance, and the more likely it is to split the wood (particularly in the case of a thick-gauge nail). When joining two pieces of timber together, the length of nail needs to be about two-and-a-half times the thickness of the attached timber. When plywood or other wood-based sheet material is being nailed to timber, the nail length should be four times the thickness of the plywood.

So called 'improved nails' (CP 112: Part 2: 1971), such as annular ring shank or square twisted nails, have greater resistance to withdrawal and lateral loading than ordinary plain wire nails. Cement-coated or resin-coated nails also have greater pull-out strength. In the case of the latter, the friction heat of driving the nails softens the resin, which later sets to adhere the nail shank to the base. Improved and coated nails discourage the tendency of

73

Table 4.1 Types and uses of wood nails

	Nail type	Head style
Round plain wire (French nail)	Flat circular
Purlin	Flat circular
Clout (slate nail)	Flat circular
Plasterboard	Circular countersunk
Square twisted (drive screw)	Flat circular
Annular ring-shanked	Flat circular
Machine driven	Flat circular semi-circular T-shaped
Lost head	Small oval
Panel pin Veneers pin	Small round

Length range (mm)	Thickness range (mm)	Material	Applications and notes
15–200	1.4–80	Mild steel	Structural timber and general carpentry applications
150–250	3.7	Copper	A round plain wire nail used where ferrous nail might corrode
10–100	2–4.5	Mild steel copper aluminium	A large-headed version of a round plain wire nail for fixing roofing felt, slates, plaster-board, etc.
31–38	2.6	Mild steel	A nail with a serrated shank for fixing plasterboard
20–200	2–8	Mild steel aluminium silicon bronze	Improved nail (CP 112: Part 2: 1971) for structural timber joints
20–200	2–8	Mild steel aluminium silicon bronze	Improved nail (CP 112: Part 2: 1971) for structural timber joints
30–100 in strips or coils	2.2–3.1	High tensile steel aluminium	Mostly operated by pneumatic machines to fabricate structural timber assemblies
15–75	1–3.75	Mild steel copper aluminium	Wire nail, often with an oval shank, for use in visible locations, punched home and the hole filled
15–25	1–2.6	Mild steel copper aluminium	Wire nail for fixing light timber components; easily punched home and hole filled. A finer version of this nail is called a veneer pin

	Nail type	Head style
	Hardboard pin	Diamond profile
	Floor brad	Rectangular
	Serrated flooring nail	T-shaped
	Sprig	Headless
	Escutcheon pin	Dome
	Tack	Flat circular
	Large-headed clout nail	Flat circular
	Tile peg	Circular countersunk
	Corrugated sheeting nail	Dome or spring
	Duplex headed (shutter nail)	Double heads
	Staple (equal or unequal legs)	–

Length range (mm)	Thickness range (mm)	Material	Applications and notes
20–25	1.6–1.4	Mild steel	Wire nail for use with hardboard; no punching home required, merely filling of the hole
40–75	2.3–3.3	Steel aluminium	A cut nail for fixing softwood flooring
50	–	Steel	For use in a manual nailing machine for fixing softwood flooring
10–25	1.4–1.6	Steel copper	A rectangular section cut nail for fixing glazing
5–40	0.9–2.0	Steel, brass	For fixing minor ironmongery
–	–	Blued steel	For fixing fabrics or carpets
12.7–38.0	2.9–3.3	Steel copper aluminium	For fixing roofing felt
–	–	–	For fixing roof tiles
25.4–76.2	3.3–5.9		For fixing corrugated roof sheeting
45–100	3.0–5.6	Mild steel	A wire nail for use in situations where withdrawal is anticipated.
3.0–75	Wide range	Steel	Hand- or machine-driven for many strengths of joint from constructional to tacking insulation in place

nails to work out during loading movement. This characteristic is known as 'popping'. Indented shank nails (serrated-edge floor nails or plasterboard nails) also have increased resistance to 'popping'.

More information on the structural performance of nails is given in the background section of this chapter.

Appearance required

Generally a nail joint is not chosen for its appearance. Therefore nails in visible positions are usually lost-head wire nails (or panel pins in lighter applications), punched below the surface and the nail hole filled. In damp locations, corrosion staining can result from the use of mild steel nails. Even nails with a corrosion-resistant coating may be subject to this defect if hammering damages the coating. Non-ferrous nails should be used in these situations. Some specialist nails, like escutcheon pins, are designed with dome heads and are intended to be seen.

Corrosion hazard

Rusting of ferrous nails can occur in damp locations, as explained above, even when they are galvanized if the coating is damaged. Copper or aluminium alloy nails should be used in these applications. Comments made in Chapter 2 concerning contact corrosion should be borne in mind if two metals are to be used together.

Copper nails should not be used with cast iron, mild steel, galvanized steel, zinc or aluminium. Mild steel nails should not be used with aluminium, zinc or galvanized steel.

Setting instructions

Nails are one of the simplest forms of mechanical fixing device and a step-by-step description of how to drive a nail is unnecessary. However, the following points should be borne in mind:

1. If a thin timber component is to be fixed, an oval nail of small shank size is less likely to split the wood so long as the axis of the oval is aligned with the grain of the wood. Blunting the nail point can help to avoid splitting very small timber components.
2. Pilot holes, 80 per cent of the nail shank diameter, should be used when nailing hard hardwoods.
3. Firm blows of the hammer should be applied to the nail head so that the impact is precisely aligned to the shank of

the nail. Do not hit the nail too hard until it has penetrated a reasonable distance into the base this avoids initially skewing the nail and later bending its shank.

4. Avoid defacing the surface of the component, particularly if it is to remain visible, with blows of the hammer head. Use a nail punch to set lost-head nails below the surface of wooden subsidiary components and fill the nail hole.

Note: The techniques for driving a nail were set out about 250 years ago in the 1736 edition of the *City and County Purchaser and Builder's Dictionary* in the following words:
There is requir'd a pretty Skill in driving a Nail, for if, when you set the Point of a Nail, you be not curious in observing to strike the flat Face of the Hammer perpendicularly down upon the Perpendicular of the Shank, the Nail, unless it have good Entrance, will start aside, or bow, or break, and then you will be forced to draw it out again; therefore, when you buy a Hammer, chuse one with a true flat Face.

In this respect, nothing very much has changed in the last 250 years!

A2 Woodscrews

The driving action of woodscrews tends to clamp the subsidiary component to the timber base more tightly than the driving of nails. As a result a screw joint has greater strength than a nailed or stapled joint. Screws have superior withdrawal resistance and strength in lateral load (shear). Diameter for diameter, a nail needs a greater shank length to equal the shear performance of a woodscrew.

Screws also have the additional advantage of being capable of removal and replacement if subsequent adjustment of the assembly is necessary.

Screws, however, are more expensive in themselves and are more costly to install than a nail. They usually require the drilling of a pilot hole before being driven and, in certain circumstances, a clearance hole through the subsidiary component.

Driving can be undertaken manually or using an electric or pneumatic screwdriver (Figs. 4.2, 4.3).

Screws of a type have been used since Roman times, but the first recorded patent for machine-made screws was granted in 1760 and screws (albeit crude by today's standards) began to be

(a)

(b)

Fig. 4.2 (a) BIF Holz-Her power screwdriver (b) Duo-fast
circular magazine screwdriver

mass-produced shortly after this date. There is now a number of types of woodscrew available with varying head styles, drive profiles, thread designs and material of manufacture.

Description A woodscrew consists of a head, containing a driving profile to engage with the screwdriver, and a pointed shank with a raised spiral thread running from the point throughout the majority of the shank (Fig. 4.3). The screw thread cuts a matching groove in the timber to make a firm anchorage for itself. It needs a pre-drilled pilot hole for effective placing and is driven through the subsidiary component (with or without a clearance hole) into the base in one fixing operation.

Fig. 4.3 GKN woodscrew

Head styles include countersunk, raised countersunk and round, containing a slotted, recessed or clutch driving profile (Fig. 4.4). The thread is usually a single spiral and is formed by recessing a groove in the shank of the screw. In this case the diameter of the thread is the same as the diameter of the shank: the shank is said to be *unrelieved*. Recently double-spiral threads have been introduced in which the thread is set above the shank diameter. The shank, in this case, is said to be *relieved* (Fig. 4.5).

Woodscrew diameters are still given in the traditional screw gauges from 0 to 20. (Unlike the standard wire gauge, the larger the number, the thicker the screw). Lengths are now quoted in millimetres.

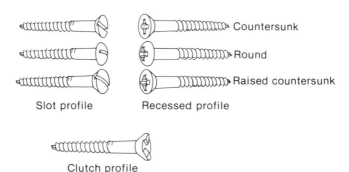

Slot profile   Recessed profile

Countersunk
Round
Raised countersunk

Clutch profile

Fig. 4.4 Head styles and drive profiles in woodscrews

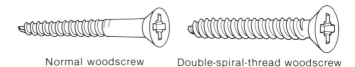

Normal woodscrew   Double-spiral-thread woodscrew

Fig. 4.5 Twinfast screw from GKN with double spiral thread

Screws are supplied in a number of different metals and finishes. Table 4.2 gives an example of GKN's range of screw types and sizes.

Applications Woodscrews are used to make joints of varying structural performance between pieces of wood, or between a variety of subsidiary components (such as sheet material, metal fixing straps and light metal accessories) and a timber base. The selection of the right screw for the job will be influenced by: (a) the performance required of the fixing; (b) the ease of driving the screw; (c) the appearance required; (d) the likelihood of corrosion.

Table 4.2 One manufacturer's range of woodscrew sizes and types

Material	Head/ drive profile	Gauge no.	Lengths (mm)	Thread type
Steel	Countersunk slot	0–20	6.4–152.4	Single
Steel	Round slot	0–14	6.4–76.2	Single
Steel	Raised countersunk slot	4–10	9.5–50.8	Single
Steel	Countersunk recessed	3–14	9.5–76.2	Single
Steel	Round recessed	3–12	9.5–50.8	Single
Steel	Raised countersunk recessed	4–10	12.7–50.8	Single
Steel	Countersunk recessed	3–12	12.7–63.5	Double
Steel	Round recessed	4–10	9.5–50.8	Double
Steel	Raised countersunk recessed	4–10	9.5–50.8	Double

Material	Head/ drive profile	Gauge no.	Lengths (mm)	Thread type
Steel	Countersunk clutch	6 – 12	19.05 – 50.8	Single
Brass	Countersunk slot	0 – 20	6.4 – 101.6	Single
Brass	Round slot	1 – 14	6.4 – 50.8	Single
Brass	Raised countersunk slot	3 – 12	9.5 – 50.8	Single
Brass	Countersunk recessed	4 – 12	9.5 – 44.5	Single
Brass	Round recessed	4 – 12	9.5 – 44.5	Single
Brass	Raised countersunk recessed	4 – 12	9.5 – 44.5	Single
Stainless steel	Countersunk slot	4 – 14	12.7 – 76.2	Single
Stainless steel	Round slot	4 – 14	12.7 – 76.2	Single
Stainless steel	Raised countersunk slot	4 – 14	12.7 – 76.2	Single
Stainless steel	Countersunk recessed	4 – 10	15.8 – 50.8	Single
Stainless steel	Round recessed	4 – 10	15.8 – 50.8	Single
Stainless steel	Countersunk recessed	4 – 10	15.8 – 50.8	Double
Stainless steel	Round recessed	4 – 10	15.8 – 50.8	Double
Stainless steel	Raised countersunk recessed	4 – 10	15.8 – 50.8	Double
Aluminium alloy	Countersunk slot	4 – 12	12.7 – 50.8	Single
Aluminium alloy	Round slot	4 – 12	12.7 – 50.8	Single
Aluminium alloy	Raised countersunk slot	4 – 12	12.7 – 50.8	Single
Aluminium alloy	Countersunk recessed	4 – 10	12.7 – 31.7	Single
Aluminium alloy	Round recessed	4 – 10	12.7 – 31.7	Single
Aluminium alloy	Raised countersunk recessed	4 – 10	12.7 – 31.7	Single
Silicon bronze	Countersunk slot	8 – 14	25.4 – 76.2	Single

Note: Steel and brass screws are available in a variety of finishes

Performance required of the fixing

A screw's holding power depends on the length of penetration and the gauge of the screw. Double-spiral threaded screws, such as Twinfast screws from GKN, offer a 3 per cent greater grip than conventional single-spiral threaded screws, assuming an equivalent length of thread penetration.

Unlike ordinary screws, 60 per cent of whose shanks are threaded, the Twinfast shank is more completely threaded. In fact, shorter Twinfast screws (up to 19 mm long) are threaded close to the underside of the head, offering a 25 per cent increase in holding power and making them particularly suitable for fixing hinges to chipboard bases and similar applications. Double-spiral threaded screws longer than 32 mm are threaded for three-quarters of their length. Double-threaded screws have enhanced performance when fixing to low-density chipboards, blockboards or softwood. What is more, the fact that the thread diameter is greater than the unthreaded shank (i.e. the shank is 'relieved') avoids the wedging action of an ordinary screw and minimizes the danger of the wood splitting.

Ease of driving

The ease with which a screw can be driven can be improved, particularly in positions of difficult access, by selecting a recessed driving profile (Phillips or Supadriv) rather than a slot. Supadriv, particularly, has been developed by GKN to improve the driving of screws at an angle. With slot-drive profiles and other recessed patterns, the driver may ride out of the recess (cam out) if the screw and the driver are not in perfect alignment. Supadriv, incidentally, is an improved version of this company's Pozidriv recessed head.

Generally, recessed driving profiles assist driving to higher torques, they aid removal and reduce recess damage from badly adjusted clutches on power drivers.

Clutch-headed screws are only used where later removal of the screw needs to be positively prohibited (e.g. in places where vandalism can be anticipated). Clutch driving profiles are normal slot profiles from which the anti-clockwise face of the slot has been chamfered off.

Double-spiral threaded screws tend to drive more quickly than single-thread screws. Their gimlet points also function rather like drill points, resulting in balanced driving – another aid to ease of placing.

84

Appearance required

Head styles can be chosen for the appearance required – countersunk, raised countersunk or round. Also the material of manufacture or finish can be chosen to match the metal of the component being fixed (dark Florentine bronze, barrel chromium plated, light bronze metal antique, etc.). Plastic snap-on caps or push-fit or screw-in dome caps are available to cover the screw head for certain special applications.

Countersunk headed screws are used for general joinery work or when the component has been prepared with sinkings to receive them. In timber-to-timber applications, where removal is anticipated, countersunk screws are often used with screw cups (Fig. 4.6). Raised countersunk-headed screws are normally used with high-standard ironmongery, the driver blade being less likely to damage the component. Round-headed screws are used especially (but not exclusively) for fixing general ironmongery or metal plates which are too thin to receive countersunk screws.

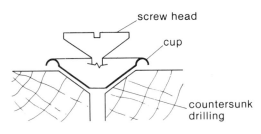

Fig. 4.6 Screw cups

Corrosion

There is a number of different types of woodscrew available for use in corrosive environments. Finishes applied to ferrous screws are applied usually for reasons of appearance, rather than corrosive protection. For conditions where rusting could take place, non-ferrous and corrosive-resistant screws should be used – aluminium alloy, brass, silicon bronze or stainless steel. Plated or coated-steel screws, such as bright-zinc plated, sherardized and black-japanned screws, should only be used externally when painted. Other plating, such as nickel, cadmium, barrel chromium and the various shades of bronze coating, should only be used internally.

Comments in Chapter 2 regarding contact corrosion should be borne in mind if two different metals are being used in close contact.

Setting instructions

1. Drill a pilot hole through the component and into the base. This is required for all but the smallest gauges of screw. In hardwood and dense chipboard the diameter of the pilot hole should be 90 per cent of the screw diameter; in softwoods and low-density chipboards, 70 per cent of the screw diameter.
2. The pilot hole should be shorter than the penetration depth of the screw from 3 mm for screw gauges 3 and 4, up to 9 mm for gauge 14.
3. Insert the screw into the pilot hole in the component and drive into the base.

Note: Where double-spiral thread screws are concerned, the component should be drilled with a clearance hole, the same diameter as the screw. This allows the component to be pulled tightly down on to the base by the tightening of the screw.

When power tools are used to drive screws, pre-drilling pilot holes is often unnecessary.

In order to avoid splitting the wood, screws should be no nearer than 10 times the diameter of the screw to the end of the component along the grain and 5 times the diameter of the screw across the grain. The first dimension can be reduced if the pilot hole is correctly drilled and both dimensions can be reduced if double-thread screws are used. Centre-to-centre spacing of screws should be 10 diameters along the grain and 3 diameters across.

Specialist screws

There is a number of other devices, developed from the wood screw, which are used for special applications.

Dowel screws These are steel dowels, threaded at both ends, which are used to make end-to-end secret connections between wood components (Fig. 4.7). The size range is limited; usually from screw gauge 6 to 12 and in lengths 25 to 50 mm.

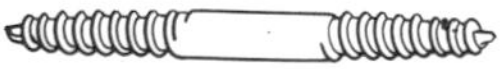

Fig. 4.7 Dowel screw

Headless woodscrews These steel devices are in effect dowels, threaded at one end and cut square at the other. The square end is given a slot driving profile. The range is extremely small; usually restricted to 12 gauge diameter and lengths of 25 and 37 mm (Fig. 4.8).

Fig. 4.8 Headless screw

Chipboard joinery screws These are one-piece fasteners for making strong joints in chipboard or timber boards meeting at right angles. They are available in 5 and 7 mm diameters and lengths of 40 and 50 mm respectively. They have a double spiral thread.

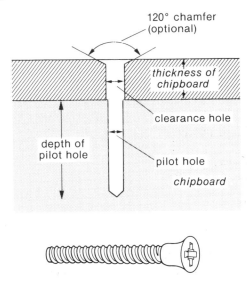

Fig. 4.9 Chipboard joinery screw and drilling data

Setting instructions

1. Select the correct size of screw for the thickness of board being joined – 5 mm screw for 12 mm chipboard and 7 mm screw for 16 mm chipboard.
2. Drill a clearance hole, the diameter of the screw, through the component.
3. Drill a pilot hole in the base, 80 per cent of the diameter of the screw and slightly deeper than the penetration of the screw.
4. Insert the screw and drive home.

A3 Coach screws

Unlike woodscrews, coach screws are not driven by a screw-driver. In addition they do not need a pilot hole. They are hammered into the wood and then tightened with a spanner.

Description Coach screws are medium- to heavy-duty threaded devices manufactured in mild steel in a limited range of sizes from 6 to 25 mm diameter and in lengths up to 400 mm (in the UK the maximum diameter obtainable can be as little as 12.7 mm, and length 150 mm). Each has a square head with which a spanner can engage.

Applications Coach screws are intended to make medium- to heavy-duty timber-to-timber, or steel-to-timber connections in which the structural performance is more important than the appearance. They are quick to install, not always requiring a pre-drilled, matching hole in the component and the base. Coach screws are removable by means of a spanner. They can be re-placed in the same hole, but repeated replacement invariably means loss of holding power.

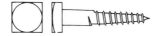

Fig. 4.10 Coach screw

Setting instructions

1. Hammer the coach screw into the timber. If the subsidiary component is metal, this will have to be pre-drilled. It is sometimes more convenient to drill a starter hole through the timber subsidiary component. This makes driving more controlled when the component is substantial; and if less robust, avoids splitting.
2. Finally tighten the coach screw using a spanner.

Note: To obtain the best performance from a coach screw, the plain (unthreaded) shank should penetrate the base by at least 25 per cent of its length.

A4 Nail plates

These devices are used to make edge-to-edge joints between pieces of timber, usually of the same thickness, lying in the same plane. There are two basic types: toothed metal nail plates and hand-nail plates.

The former consists of metal plates with teeth punched out of their surface in a regular pattern (Fig. 4.11). They cannot be installed by hand, but are intended to be placed by special machinery in a factory, the plates being positioned over the junction between two or more timber components and then pressed home. They are used particularly in the manufacture of trussed rafters and similar prefabricated timber assemblies.

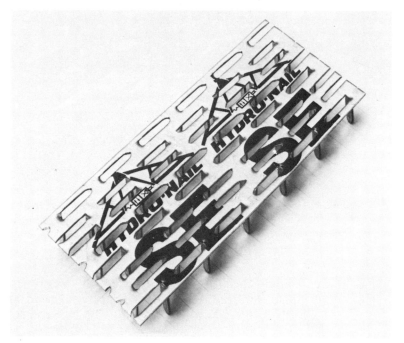

Fig. 4.11 Punched nail plate (Hydro-Air)

Because they are used by fabricators holding licences from the plate manufacturers, these devices are beyond the scope of this book.

Hand-nail plates, on the other hand, are not subject to the same restrictions. They are metal plates punched out with a pattern of nail holes which are intended to receive normal nails, driven either by hammer, or by a nailing machine (Fig. 4.12). They are used to make similar edge-to-edge joints as the punched nail plate.

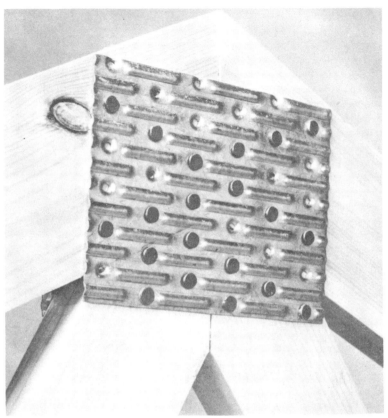

Fig. 4.12 Nail plate (Bat-U-Nail truss plate)

The most commonly specified nails used with hand-nail plates are 30 × 3.75 mm sherardized twist nails, hand-driven, or 30 × 3.35 mm round-shank T-nails driven by a pneumatic nailing machine.

It should be noted that hand-nail plates are similar to the metal accessories examined in section C of this chapter and could equally well have been considered in that section.

Description Hand-nail plates are usually made of 1 mm galvanized mild steel, perforated with rows of holes through which the specified size of nails is to be driven. The range of plate sizes available from different manufacturers varies considerably. The most commonly-used sizes are in a range from approximately 100 to 300 mm. Exact sizes are selected according to the size and type of joint to be made and the number of nails that must be driven into each piece of timber.

Applications Nail plates can be used to construct edge-to-edge joints between two or more pieces of timber of the same thickness and lying in the same plane. They are particularly useful in the fabrication of such assemblies as trussed rafters, where their use allows smaller sections of timber to be joined together without resort to large plywood gusset plates. Nail plates are applied to both sides of the joint. The nail plate manufacturers' literature gives details of the size of plate and the number and size of nails to be used in various applications.

Setting instructions

1. Ensure that the pieces of timber to be joined have accurately been cut to fit and are firmly clamped together with no gaps between their contact surfaces.
2. Select a nail plate of the correct size and place it over the junction of the timber components in such a way as to allow the required number of nails (specified by the manufacturer or the engineer) to be driven into each piece of timber.
3. Distribute the nails evenly over the surface of the plate avoiding driving nails too close to the edge of the timber (usually minimum edge distance can be taken as 27 mm and end distance 38 mm).

A5 Corrugated fasteners

These fasteners, often referred to as 'dogs', are the simplest and most lightweight of the fasteners which are used to make edge-to-edge connections between pieces of timber in the same plane – butt joints and mitres (Fig. 4.13).

Description A corrugated fastener, as its name suggests, comprises a small corrugated plate with one edge sharpened for driving into the timber. The corrugations are skewed towards one another as they approach the sharpened edge so that the action of driving the fastener draws the parts of the joint together. Various widths of fastener are obtainable and depths range from 6 to 25 mm.

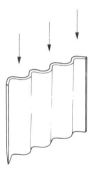

Fig. 4.13 Corrugated fastener

Applications Corrugated fasteners are used to make light-weight connections between timber components in the same plane, the device being driven across the line of the joint so that one half of the fastener is in one timber component and one half is in the other. They are only used where the appearance of the joint is not important. The slight taper on the fastener tends to draw the joint components together.

Section B: Through-fixings – type 2 (a)

Through-fixings, used to make overlap joints between timber components or between metal components and timber bases, depend for their effectiveness on some form of bolt to clamp the parts of the joint together. This clearly, therefore, involves the pre-drilling of an accurate hole through both component and base before the assembly of the joint.

In addition to the simple bolt (and its variants, the carriage bolt and the coach bolt), this family of fixings includes toothed connectors, split-ring connectors and shear plate connectors.

B1 Bolts

Hexagonal headed bolts, used alone to make structural connections between substantial pieces of timber, or between steel plates and timber bases, do have disadvantages if heavy loading is anticipated and good structural performance is needed.

Because the bolt exerts stress only on the relatively small area of timber through which it passes, the structural capacity of the whole timber area is never fully used (maybe only 40 to 60 per cent of it is stressed). In addition, the clamping effect of tightening the bolt does not result in sufficient friction between adjoining faces of timber to enhance the strength of the joint. Seasonal moisture content changes in the wood and thermal movement in the bolt in time destroy even this limited effectiveness. As a result, a number of connections has been developed for use with bolts which improve their performance by spreading the load more widely and making component contact more positive (see sections B2, B3 and B4).

Two points should be remembered when obtaining bolts for use in timber construction:

1. Make sure the threaded portion of the shank is long enough to draw the joint together fully. This is particularly important when a toothed connector is being used (see section B2). Usually bolts, because they are designed primarily for metal construction, are only threaded up the shank for a distance of two-and-a-half times the bolt diameter, but bolts with longer threads can be obtained by special order.

2. Over-sized washers are usually essential under both the bolt head and the nut. This is to avoid the washer digging into the wood if loading tends to cause angular displacement of the bolt (Fig. 4.14). Once more the standard sized washers supplied with the bolt tend to be satisfactory for steel joints, but not for wood.

Fig. 4.14   Angular displacement of a bolt joint

Carriage bolts have practically disappeared from the British market. They have square heads and nuts, coarse buttress threads and are now only met in alteration work. Coach bolts, on the other hand, are still used to some extent in the building industry. Carriage bolts are sometimes confused with coach bolts, maybe because the Americans refer to coach bolts as carriage bolts.

The coach bolt has a cup head with a square underhead. This is

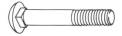

Fig. 4.15 Coach bolt

inset into the wood and resists the tendency of the bolt to revolve
when the nut is being tightened. Coach bolts are ideal where access for a spanner is only available on one side of a joint.

Description Most bolts used in timber connections are
ordinary mild-steel bolts (black bolts) with hexagonal heads and
nuts. They are available in a wide range of lengths and diameters
which are detailed more fully in Chapter 5.

Applications Bolts are used without the aid of an
additional connector, to make medium-duty timber-to-timber or
metal-to-timber connections. They normally have to be fitted
with over-sized washers to spread the load under the bolt head
and the nut in order to avoid timber deformation.

Setting instructions

1. Drill a hole through the component and the base, 1.5 mm
   greater than the bolt diameter for bolts up to 25 mm diameter. (A greater allowance will be needed for larger
   bolts.) Drilling can be carried out through component and
   base in one operation by spiking the members of the joint
   together, if this is convenient: otherwise careful marking
   out and the possible use of a template is recommended.
2. Assemble the elements of the joint. Place the over-sized
   washer on the bolt and insert in the drilling.
3. Place a second over-sized washer on the threaded end of
   the bolt and run-on the nut. Tighten with two spanners;
   one applied to the head to avoid it revolving, the other
   used to turn the nut.

B2 Toothed connectors

Toothed connectors can be either single- or double-sided: the
single-sided version is used to make connections between metal
components and timber bases (or between a timber component
and a concrete base); the double-sided connector makes a permanent overlapping timber-to-timber joint. A demountable timber-

to-timber joint can be made by using two single-sided toothed connectors, back to back (Fig. 4.16).

Fig. 4.16 Toothed connectors, single and double

Description Toothed connector plates are made of galvanized steel plate and are usually available in four sizes, either round or square; 38, 51, 63 and 76 mm. There is a less common larger-sized connector (89 mm), but this is not referred to in CP 112: Part 2: 1971. All connectors but the largest and the smallest, are designed for use with an M12 hexagonal-headed black bolt; the 38 mm connector is sometimes used with an M9 bolt and the 89 mm with an M20 bolt.

Applications Connectors are designed to make reliable heavy-duty timber-to-timber connections, or joints between a steel component and a timber base. In the former case a double-sided connector is used (unless the joint is to be demountable, when two single-sided connectors, back-to-back, are installed). In the latter case a single-sided connector is used with its teeth embedded in the timber.

Setting instructions

Double-sided toothed connector (single-sided connector similar)

1. Place the timber members in the position they will assume when the joint is assembled and clamp or spike them together.
2. Drill through all timbers at the point where their centre-lines intersect. Ensure that the bolt hole (no larger than 1.6 mm greater than the bolt diameter) is drilled at right angles to the surface of the joint. (Alternatively, each member can be drilled separately using a template. Complete accuracy is essential.)

3. Position the connector between the members of the joint and centrally over the drilling. Reposition the joint components.
4. Embed the connector by either:
    (a) threading the bolt through the joint and tightening the nut to compress the gaps between components (in the case of hard timbers, this may place too great a strain on the threads of the bolt or nut);
    (b) passing a high-tensile steel screwed rod with large plate washers through the joint and tightening the nuts (compression tools like this are obtainable from the fixing manufacturer);
    (c) using screw cramps or hydraulic pressure.
5. Once the teeth of the connector are fully embedded, remove the compression tool (if one were used) and replace it with the black bolt with square washers, and tighten.

Note: In positioning the connector, specified minimum edge and end distances should be observed carefully. The teeth of the connector should be completely embedded, but no effort should be made to squash the plate of the connector into the wood. The components of a correctly-made toothed connector joint will stand apart by the thickness of the connector plate.

Where more than three connectors are used in any one joint, a threaded rod compression tool must be used to achieve proper embedment.

B3 Split-ring connectors

Split-ring connectors are used to make overlapping timber-to-timber joints which will carry greater loads than toothed connector joints.

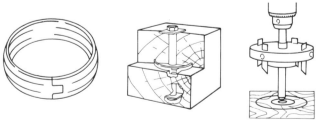

Fig. 4.17 Split-ring connector and grooving tool (also used with shear plate connectors)

Description Split-ring connectors consist of a galvanized hot-rolled low-carbon steel split ring, either bevelled or straight-sided, which is used in conjunction with a hexagonal-headed black bolt. Ring diameters are 64 and 102 mm (for use with M12 and M20 bolts respectively). At least one manufacturer produces a 50 mm diameter split ring, but this size is not included in CP 112: Part 2: 1971.

Applications The split ring is used to make heavy-duty timber-to-timber fixings of greater strength than a double-sided toothed connector. The bevelled ring is easier to place and has less slip under load than the straight-sided ring. Both types are set in a circular groove, cut in the contact faces of both pieces of timber, at the centre of which groove is the drilling for the bolt which holds the joint together. The adjoining faces of the timber components are in contact in a correctly assembled split-ring joint, not standing slightly apart, as in the case of the toothed connector joint.

Setting instructions

1. Place the timber members in the position they will assume when the joint is assembled and clamp or spike them together.
2. Drill through all timbers at the point where their centre-lines intersect. Ensure that the bolt hole (no larger than 1.6 mm greater than the bolt diameter) is drilled at right angles to the surface of the joint. (Alternatively, each member can be drilled separately using a template. Complete accuracy is essential.)
3. The contact surfaces should be grooved with a grooving tool to a depth of half the ring concentrically about the bolt hole. There are various tools available to carry out this work, some obtainable from the fixings manufacturers. A cutter-head with a 'pilot' which can be located in the bolt hole is one option; or the bolt hole and the groove can be cut in one operation using a power tool. There is, however, a limit to the diameter of ring groove that can be cut by this type of tool. Grooves should be cut slightly larger than the diameter of the split ring to ensure its best performance, even should the timber dry and shrink during use (see the note below).

4. Remove all chips and shavings from the groove.
5. Expand the ring and insert it in the grooves in both timber members.
6. With the joint reassembled, insert the bolt, complete with washers, run-on the nut and tighten.

Note: The dimensions of the groove to receive the split ring should always be those recommended by the manufacturer. In the case of 63 and 100 mm diameter split rings one manufacturer recommends the dimensions given in Table 4.3. The observance of edge and end minimum distances is essential.

Table 4.3

Split ring diameter	63 mm	100 mm
Inside diameter of groove (mm)	65.0	104.0
Width of groove (mm)	4.6	5.3
Depth of groove (mm)	9.5	12.7

B4 Shear plate connectors

When heavy-duty timber-to-steel joints are required, the shear plate connector has an equivalent performance to a split-ring connector (i.e. better than would be obtained by using a single-sided toothed connector).

Fig. 4.18 Shear plate connector

Description Shear plate connectors are similar to single-sided toothed connectors, except that the circular plate has a flange on its circumference rather than a series of teeth. This is inset into a groove in the timber like a split-ring connector. The circular plate is also inset into the wood. Shear plate connectors are made from pressed steel or malleable iron and are used in conjunction with hexagonal-headed black bolts. Two sizes of shear plate are commonly available: 67 and 102 mm diameter; both are used with M20 bolts.

Application Shear plate connectors make reliable heavy-duty connections between timber bases and metal components (they can also be used in pairs, back-to-back, to produce demountable timber-to-timber connections). Shear plate joints are stronger than those produced by toothed connectors. Because the shear plates are completely set in sinkings in the timber, the adjoining faces of the joint members are in contact when a correct shear plate connector joint has been made.

Setting instructions

1. Drill the timber member to receive the bolt. The drilling diameter should be no larger than 1.6 mm greater than the bolt diameter. Ensure the bolt hole is drilled at right angles to the surface of the joint. Complete accuracy is essential.
2. The timber contact surface should be grooved and rebated to receive the shear plate, concentrically about the bolt hole, using a 'dapping tool' (see note below). These are often obtainable from the fixings manufacturer.
3. Remove all chips and shavings from the groove and sinking.
4. Insert the shear plate. Assemble the joint and insert the bolt, complete with washers, and tighten the nut.

Note: The dimensions for the groove and sinking recommended by the manufacturer should be followed. In the case of 67 and 100 mm diameter shear plate connectors, one manufacturer recommends the dimensions given in Table 4.4. The observance of edge and end minimum distances is essential.

Table 4.4

Shear plate connector diameter	67 mm	100 mm
Sinking diameter (mm)	67.0	100.0
Sinking depth, maximum (mm)	11.5	16.5
Sinking depth, minimum (mm)	6.5	7.0
Width of outer groove (mm)	5.0	6.5

Section C: Fixing accessories

A number of metal fixing accessories has been developed specifically to eliminate the need to construct on site traditional joinery joints which require considerable skill and time to form. When used in conjunction with the specified nails, these metal accessories make heavy-duty structural connections between timber members meeting at right angles or in the same plane.

Joist hangers (similar to the joist hangers discussed in section D of Chapter 3, but which are nailed to timber members rather than bolted or nailed to, or built into, mass walling) remove the need to form a tusk tenon joint between a trimming joist and a trimmer joist or a notched joint between a trimmed joist and a trimmer (Fig. 4.19). Similarly, framing anchors can be used to make simple end-to-edge joints between timber framing members and cantilever brackets to extend a timber member rather than forming a spliced joint. Some of these devices are illustrated in Fig. 4.20.

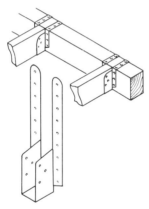

Fig. 4.19 Joist hanger to make timber-to-timber connections

The effectiveness of all these accessories depends on the thickness and length of nails used to fix them to the timber members. The manufacturer's recommendations should always be followed.

Joist hangers

Description Joist hangers are produced in a variety of sizes to suit various cross-sectional sizes of joist. They are usually manufactured of hot-dipped galvanized mild steel, from 2.7 to

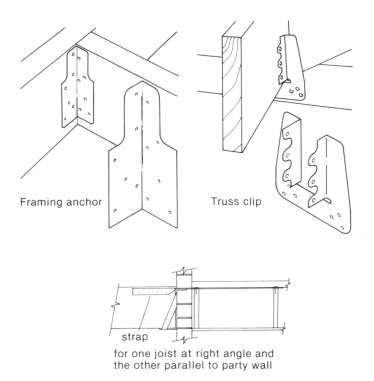

Framing anchor

Truss clip

strap

for one joist at right angle and
the other parallel to party wall

Fig. 4.20 Framing anchors, truss clips and straps

1 mm thick, depending on the strength required of the connection. These devices are usually pre-holed for fixing with 32 mm long × 3.75 mm square twisted nails, although some of the more heavy-duty hangers are intended to be used with 12 mm diameter bolts.

Applications These devices are intended to make connections between timber joists meeting at right angles. Nails should be driven through all holes in the hanger.

Framing anchors

Description Framing anchors are usually made from 1.2 mm galvanized steel and are intended to make medium-duty connections when fixed with 32 mm long × 3.75 mm square twisted nails.

Applications Framing anchors are intended to make connections between framing timbers meeting at right angles (see Fig. 4.20). Nails should be driven through all holes in the anchor.

Truss clips

Description Truss clips are usually made from 1 mm galvanized steel and are fixed with 32 mm long × 3.75 mm square twisted nails.

Applications Truss clips are used to fix a roof truss to the wall plate (see Fig. 4.20). In order to achieve the correct resistance to uplift, nails should be driven through all holes in the clip.

Straps

Description Straps are usually made from 2.5 mm thick × 35 mm wide galvanized steel, pre-punched with holes at 25 mm offset centres and are fixed with 32 mm long × 3.75 mm square twisted nails.

Applications Straps are used for a number of bracing applications, often in conjunction with built-in double joist-hangers (see Fig. 4.20). Nails should be driven through all holes backed by the timber members.

Cantilever brackets

Description Cantilever brackets are usually made from 1.5 mm thick zinc-coated mild steel to suit a range of joist widths from 38 to 75 mm and joist depths from 150 to 255 mm. Nails to be used are 32 mm long × 3.75 mm square twisted nails.

Applications These brackets can be used to form a lengthening joint between joists, provided the joint is made not too far away from a support. Manufacturers lay down precise instructions as to how the permitted dimension between joint and support can be determined. Cantilever brackets *do not* allow the indiscriminate use of random joist lengths, but they do avoid the overlapping of end-on joists over supporting walls with a consequent saving in

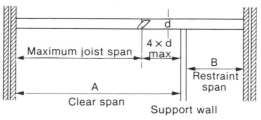

B   Normal min. length 7 times joist depth
    (d). Absolute min. length 4 times joist
    depth when fully built in or trimmed to
    adequate support using "BAT" joist hangers
    and 10 No.9 SWG. × 32 mm lg. nailed fixing.

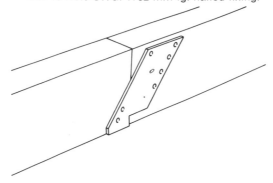

Fig. 4.21   Cantilever bracket

timber. They also result in a reduction in the cross-sectional area
of timber required. (Fig. 4.21)

Setting instructions

1. Nail the cantilever bracket to the joist to be cantilevered
    over the supporting wall. This is best done on the
    ground. The joist must be located squarely against the up-
    ward projecting lug in the base of the bracket and with
    the underside of the joist and bracket in line.
2. Notch out the underside of the other joist (the meeting
    joist) to receive the bracket.
3. Place the cantilever joist in position.
4. Place the meeting joist so that its end fits squarely against
    the upward projecting lug in the base of the bracket and
    nail in position.
5. All nail holes should be filled.

Note: Since the increased popularity of timber-framed forms of house construction, a number of specialized metal accessories have been introduced (see *Timber Frame Housing* by Jim Buchell in this Site Practice series).

Section D: Background

Strength of mechanical fixings in timber bases

The strength of mechanical fixings to timber bases is affected by the species of timber in question, some species being considerably stronger than others. The terms 'softwood' and 'hardwood' should not be confused with an indication of the strength of the wood. The distinction is purely botanical and many hardwoods are not as strong as some softwoods. Softwoods are broadly timber from coniferous trees and hardwood from deciduous trees. Most timber used for structural framing comes from the softwood category. Generally the more dense the timber, the higher are its strength properties. This is illustrated by the statistics in Table 4.5.

Table 4.5

Species	Density (kg/m^3)	Maximum bending strength (N/mm^2)	Maximum compression strength parallel to grain (N/mm^2)
Pitch pine	769	107	56.1
Douglas fir	545	93	52.1
European larch	545	92	46.7
Baltic redwood	481	83	45.0
Western hemlock	465	83	47.4
European spruce	417	72	36.5
Western red cedar	368	65	35.0

The stronger the wood, generally the better the fixing, provided of course that the quality of the pieces of timber being compared is equivalent (same disposition of knots and slope of grain, etc.) and their moisture contents are the same. All the figures in Table 4.5 relate to samples of the species with a 12 per cent moisture content.

The direction of the grain also affects the strength of the timber. For instance, because of the layout of the fibrous cells which make up the wood, compressive forces parallel to the grain can be sustained more readily than similar forces applied perpendicular to the grain. In tension, wood's strength perpendicular to the grain is lower, but parallel to the grain very high – higher, in fact, than compressive strength. All this has a marked effect on the way fixings are placed in timber joints, both in relation to the edges of the pieces of timber being joined and in relation to other fixings in multiple fixing joints.

The structural performance of mechanical fixings in timber bases, therefore, depends on the size and thickness of the timber base (and, if appropriate, the timber component) as well as the size and character of the nail, screw, bolt or connector used to make the fixing. Often the strength of the joint is determined by a combination of the characteristics of the timber member(s) and the fixing. Thus the size of the fixing is critical to the strength of the joint, but its strength can be undermined by its use in an inappropriately-sized piece of timber, or if it is placed in such a way as to cause the wood to split or form planes of weakness.

The other vital factor to be considered is the possibility of the corrosion of the metal fixing. This will be considered later in this section.

The design of joints made by timber connectors is a very specialized business and so joints containing these devices are usually provided with detailed engineering data in the form of drawings and specifications, giving precise instructions concerning the positioning of connectors. This aspect is somewhat comparable to bolted structural steelwork connections and is therefore, like them, beyond the scope of this book. Extensive recommendations are given in CP 112: Part 2: 1971 which should be consulted if specific advice is required concerning the design of joints using connectors.

There are a few pieces of basic advice which can be given concerning other structural connections using nails, staples, screws or bolts and which will help to ensure fixings of adequate performance.

Timber-to-timber joints can be loaded axially or laterally (Fig. 4.22).

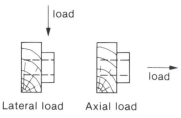

load

Lateral load   Axial load

Fig. 4.22   Lateral and axial loading

Lateral loading

All mechanical fixings achieve their greatest strength when driven into the side grain of the wood. In the case of a nail, its optimum effect is realized when the length of the nail is two-and-a-half times the thickness of the timber component. When driven into the end grain, the nail's (or screw's) performance is reduced by about a third. Also nailing or screwing into green timber or into timber that will be exposed to the weather has a reduced performance. The same applies to bolted connections.

Staples have a performance under lateral load equal to a nail one-and-a-half times the staples diameter.

In making lapped joints between pieces of timber, the nailing area is clearly limited (Fig. 4.23) and therefore the number of nails is severely restricted. To achieve the best performance it is therefore necessary to use the largest-diameter nail possible with the deepest penetration that is consistent with the sizes of the component and the base without causing the timber to split. Spacing of nails is therefore critical. This will be discussed in more detail later in this section.

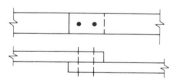

Fig. 4.23   Lap joint

If the members of the joint are in line, strong joints can be made using plywood gussets nailed with generally smaller nails (Fig. 4.24). A useful rule of thumb is that the nail length should be 4 times the thickness of the plywood – usually not less than 50 mm long. The following nail diameters (Table 4.6) are sug-

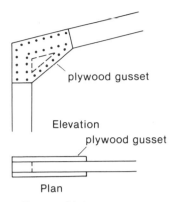

plywood gusset

Elevation

plywood gusset

Plan

Fig. 4.24 Gussetted joint

gested by TRADA (Timber Research and Development Association) Mechanical fasteners for structural timberwork, Nov 78, *TRADA Wood Information*), based on the thickness of the plywood gusset:

Table 4.6

Diameter of nail (mm)	Thickness of ply (mm)
2.64	8.0
3.25	9.5
3.66	12.7
4.47	15.8

Axial loading

This form of loading is resisted by the holding power of the fixing, usually a nail or screw (a bolt is rarely used to withstand axial loading). Normal nails are not strong withdrawal fixings, but improved nails (square twisted or annularly threaded nails) have a better performance, allowing the penetration depth of the nail to be decreased by a third without reducing holding power. A screw has considerably better performance than a nail regarding withdrawal resistance.

No type of nail or screw produces good withdrawal performance when driven into end grain, but the performance of a nail is unaffected by use in green timber or timber exposed to cyclic damp conditions. A screw, however, has reduced performance in green timber. Double skew-nailing increases resistance of nailed joints to axial failure (Fig. 4.25).

Fig. 4.25 Skew nailed joint

It should be remembered that it is the length of the thread of a screw embedded in the timber base which gives it its holding power. Threaded shank unembedded in the base has no beneficial effect on the joint.

Spacing of nails, screws and bolts

The recommended spacing of nails, screws and bolts to achieve their best performance in timber-to timber joints is set out in Table 4.7.

Corrosion

Dampness can be a cause of corrosion in ferrous timber fixings, as has already been noted in respect of other categories of ferrous fixing. However, non-ferrous alternatives are more expensive and rarely warranted except in the most critical circumstances. In positions where the fixing is to suffer damp conditions, the need for long-term, high-strength performance needs to be weighed against the additional cost of the non-ferrous alternative. Often the use of a zinc-coated fixing is sufficient defence. In the form of galvanizing it will stand a fair amount of rough treatment during installation before it is in danger of suffering premature failure.

Timber treated with water-borne preservative salts needs a stabilization period of about 14 days between treatment and the introduction of metal fixings. Otherwise fasteners *must* be non-ferrous (but *not* aluminium) after checking with the manufacturers of both the fixing and the preservative that their products are compatible. This does not apply to solvent-based preservatives.

When metal fasteners are used in timber impregnated with flame retardants, the manufacturer's advice should always be sought. It is often necessary to use non-corroding fasteners made of materials like phosphor or silicon bronze when there is the slightest danger of damp conditions when timber treated with flame retardants is being fixed.

Table 4.7 Spacing of nails, screws and bolts

Type of fixing	Illustration	Position/spacing	Relation to grain	Distance predrilled hole	Distance no hole
Nails and screws		End distance	Along	10D	20D
		Edge distance	Across	5D	5D
		Spacing	Across	3D	10D
		Spacing	Along	10D	20D
Bolts		End distance	Tension; along	7D	
		End distance	Compression; along	4D	
		Edge distance	Any; across	4D	

	End distance	Any; across	4D
	Spacing in direction of load		4D
	Spacing at right angles to load	Load across	$2.5D$ if $\dfrac{t}{D} = 1$ $5D$ if $\dfrac{t}{D} = 3$ or more (interpolate between)
	Spacing at right angles to load (bolt to edge)	Load along	$1.5D$

D = diameter of fixing; t = thickness of timber component; F → direction of load
All distances measure from centre of fixing. ← G → direction of grain

5

Mechanical fixings in metal bases

Fixings to metal bases divide broadly into two categories: heavy-duty fixings, being primarily fixings from which high and predictable performance is demanded (similar to those used to connect members of a structural steel framework together) and lightweight fixings, used generally to make connections to lighter metal bases or fixings with a less severe performance function. The first category consists of type 1 through-fixings (bolts and machine screws of various types); the second category of mainly type 2(b) base-deforming fixings.

There are two points which should be remembered in relation to both categories. Metal, as a generic description, covers a wide range of materials, some of which are more resistant to corrosion and atmospheric degrading than others. This means that the metal must either be protected from the conditions which may cause its decay, or the type of metal must be selected to be unaffected by the conditions.

Secondly, as the fixing devices are themselves made of metal and the subsidiary component may also be of metal, the problem of galvanic or bimetallic corrosion has to be considered and avoided. This problem and its avoidance has been discussed in Chapter 2 on page 9.

Section A: Heavy fixings

Heavy-duty fixings to metal bases comprise bolts and machine screws. These parallel-sided fixings are generally drilled-for, through-fixings in the category type 1 and are used in conjunction with a nut and (possibly) washers. The difference between a bolt and a machine screw is largely based on the length of its thread and there is a fair amount of size duplication between the two ranges of fixings.

A bolt is a threaded fastener with a diameter greater than 6 mm (M6) and provided with a matching nut and washers. Below that diameter, threaded devices are considered as being machine screws. But this is not a complete definition, because machine screws can be obtained up to M12 diameter (12 mm) – well within the range of bolt diameters. A machine screw is, however, threaded for the whole of its length, whereas a bolt is only threaded for a part of its shank – usually about two times its shank diameter. This again becomes confusing in the case of shorter bolts, where this formula implies an almost complete threading of the bolt shank. In practice bolts below the lengths shown in Table 5.1 are considered as being machine screws.

Table 5.1

Shank diameter (mm)	Black bolt lengths (mm)	Precision bolt lengths (mm)
M5	25	20
M6	25	25
M8	30	30
M10	35	35
M12	40	40
M16	35	45
M20	45	55
M24	55	65
M30	85	80
M36	100	90
M42	120	110

A1 Bolts

The introduction of the metric bolt has rationalized a chaotic situation of varying threads used in this country, including British Standard Whitworth, British Standard Fine, British Association, Unified Coarse and Unified Fine threads. These still may be encountered on occasions. Now the ISO metric thread has been introduced, eventually to supersede not only previous British threads, but some continental threads as well (DIN, CNM and SI metric threads). There is also an ISO inch (unified) thread, but that is also likely to fall out of use.

ISO metric bolts are always preceded by the prefix M, followed by the diameter in millimetres (M10, M12 etc.). Coarse and fine ISO metric threads are available.

Various head styles of bolts are manufactured (Fig. 5.1) but the most common is the hexagonal all head, matched by a hexagonal nut.

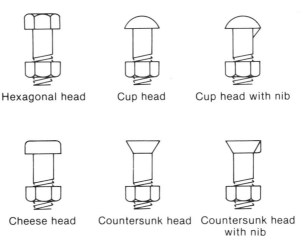

Fig. 5.1 Bolt head styles

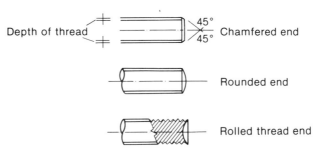

Fig. 5.2 Bolt end finishes

There are also a number of bolt end finishes (Fig. 5.2).

The effectiveness of a bolted connection is clearly dependent on the size, type and number of bolts, as well as their spacing and location. These are matters which will be detailed on the engineer's drawings or in his specification, assuming that they have structural significance, and therefore need not be covered in this book. Some guidelines on the spacing of bolts is included in section C of this chapter.

Black bolts

Description The name 'black bolt' does not refer to the appearance of the bolt, but implies an unfinished bolt with a comparatively wide range of tolerances – considerably wider than would be permissible in the case of a precision bolt. The current British Standard covers a range of diameters from M5 to M68, although manufacturers tend to stock a considerably reduced number of sizes. They are made from mild steel of a limited strength grade (see section C, p. 143) and are identifiable by an 'M' embossed on their heads, without any strength grade identification. Matching nuts and washers are supplied with the bolts.

Applications Black bolts are used for general structural steel applications where high performance is not demanded of the bolt and where slip and vibration are not critical considerations. Access is needed to both sides of the joint to make the fixing. The threaded portion of the bolt should not enter the shear plane (Fig. 5.3). To this end, bolts with shorter thread lengths, only one-and-a-half times shank diameter in length, are available.

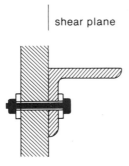

shear plane

Fig. 5.3 Shear plane

Precision bolts

Description These bolts are usually made of carbon or alloy steel, supplied heat-treated after manufacture (dull black) or bright finished. They can be supplied in stainless steel. Generally, precision bolts have a greater degree of accuracy than black bolts and their strength grade is identified on their heads (8.8, 10.9 etc. – see section C of this chapter for a description of this designation). Matching nuts and washers are supplied with the bolts.

Applications Precision bolts are used where higher performance is required of the bolted connection. Access is required to both sides of the joint and the same proviso applies regarding the intrusion of the thread into the shear plane, as was made in respect of black bolts.

High-strength friction grip (HSFG) bolts

Description High-strength friction grip bolts are usually made of carbon steel and, when tightened to a predetermined shank tension, they clamp the component to the base in such a way that the loads are transferred by friction between the parts, not by shear in, or bearing on, the bolts or parts of the connected members of the joint. Part 1 general grade bolts (BS 4395) come in a size range from M12 to M36; Part 2 bolts (higher-strength grade) from M16 to M33. Thread lengths on Part 2 bolts are slightly longer than the normal thread length. High-strength friction grip bolts are used with various types of washer, flat round, flat square and square tapered (Fig. 5.4).

Fig. 5.4 Nuts and washers for use with HSFG bolts

In the case of Part 2 bolts, the strength grade is marked on the bolt; Part 1 bolts have three radial lines 120° apart on their heads, but no strength grade.

Applications High-strength friction grip bolts are used in high-strength applications where slip must be eliminated. Once more access is required to both sides of the joint to make the fixing. Part 2 bolts should not be used where the loads operate along the axis of the bolt. Instead use a waisted shank bolt.

As it is the action of tightening the nut on the bolt which induces the tension in the shank and causes the friction between the joint plies, the procedure of tightening must be carefully con-

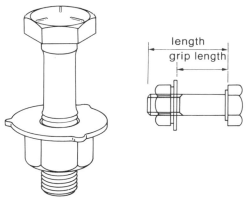

Fig. 5.5 HSFG bolt showing head markings

trolled. There are two methods: the part-turn method and the torque control method.

Setting instructions

Part-turn method (Part 1 bolts only)

1. Tighten the bolt using a podger spanner in the normal way to draw the joint surfaces into close contact.
2. Record the relative positions of the nut and protruding bolt end with a paint line or a cut from a cold chisel.
3. Finally tighten the nut using an impact wrench so that it turns relative to the bolt by the amount specified – usually half or three quarters of a turn.

Torque control method (Part 1 or Part 2 bolts)

1. Using a calibrated tightening device, tighten the nut to the torque necessary to induce the minimum bolt tension set out in the manufacturer's data. This is the preferred method.

Note: Bolts should be tightened in a staggered pattern and, in large groups of bolts, work should start from the centre of the group and move outwards. The use of a split-ring washer under the bolt head can help to maintain the pre-load of tightening. A matching plain washer should always be used under the nut.

Load-indicating bolts (LIB)

Description This is a special HSFG bolt which gives a physical indication when the bolt has been tightened correctly. It has a square head with a triangular pad under each corner of the head and a recessed section between pads.

Applications Load-indicating bolts are used in all applications where normal HSFG bolts could be used. Access is once more required on both sides of the joint. The tightening of the bolt compresses the triangular pads (A) (Fig. 5.6) and reduces the gap at C. When this reaches the minimum set down by the manufacturer, the required bolt tension has been achieved.

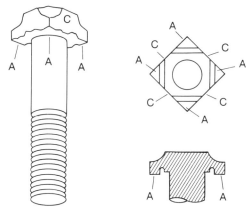

Fig. 5.6 Load indicating bolt

This is a more reliable indication of shank tension than the torque control method applied to normal HSFG bolts which tends to measure the friction between the mating surfaces of the thread, rather than shank tension.

Setting instructions

1. Tighten the bolt connection in the usual way.
2. With a feeler gauge, ascertain when the gap between the underside of the bolt head and the washer (or steel component) at C has been reduced to 1 mm (or the dimension laid down by the manufacturer). The bolt has then achieved the necessary bolt tension for the friction between the plies of the joint to have been achieved.

Note: Manufacturers usually recommend the size of no-go gauge should be 0.75 mm thick.

Load-indicating washers, which operate on similar principles, are also available for use with conventional HSFG bolts.

Waisted high-strength bolts

Description This special HSFG bolt is usually made of high-duty alloy steel, and its shank diameter is reduced (or 'waisted') below the root diameter of the thread. This allows an increased tension to be achieved in the bolt due to its high degree of elastic tension. Failure, therefore, occurs in the shank and not in the thread as is the case of standard bolts.

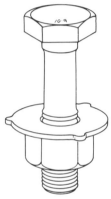

Fig. 5.7 Waisted high-strength bolt

Applications Waisted high-strength bolts are used in positions of severe fatigue or where bolts are subjected to tension as well as shear. Access is required to both sides of the joint to make connection.

Accessories

There is a number of types of locking nut available to withstand the effects of vibration on bolted connections (Fig. 5.8). These include slotted and 'castle' nuts, for use with a split pin or similar device, and self-locking turret nuts which include a turret section above the hexagonal nut which is cut twice and deformed to de-pitch the thread above the slot.

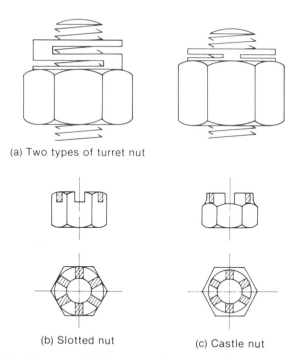

(a) Two types of turret nut

(b) Slotted nut

(c) Castle nut

Fig. 5.8 Locking nuts (a) Two types of turret nut (b) Slotted nut (c) Castle nut

A2 Machine screws

Machine screws form an uneasy bridge between heavy-duty and light-duty metal base fixings. The difference between machine screws and bolts has been explained on page 112. Generally, machine screws are used to make medium-duty fixings and never move into the upper level of heavy-duty structural fixings, as do bolts. British Standard 4183: 1967 gives a preferred range of machine screw sizes from M2 to M12 diameters with lengths from 5 to 90 mm, but the majority of machine screws in common use are in the lower, light-duty end of the range.

Description Machine screws are made in a variety of metals, both ferrous and non-ferrous, and in a number of finishes. Head styles can be pan, countersunk, raised countersunk or cheese head (Fig. 5.9) with slot or recess driving profiles. There is also a range of hexagonal socket-headed screws, including cup, button and countersunk head styles (Fig. 5.10) as well as (at the larger

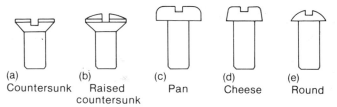

Fig. 5.9 Machine screw head styles (a) countersunk (b) Raised countersunk (c) Pan (d) Cheese (e) Round

end of the range) the hexagonal bolt head styles.

Machine screw points can vary from the normal flat end, to special lead-in ends like the cone point or dog point.

Matching nuts and washers are supplied with machine screws.

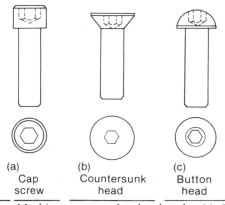

Fig. 5.10 Machine screw socket head styles (a) Cap screw (b) Countersunk head (c) Button head

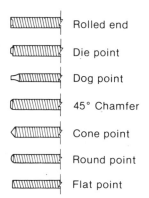 Rolled end

Die point

Dog point

45° Chamfer

Cone point

Round point

Flat point

Fig. 5.11 Machine screw points

Applications Machine screws are used extensively to connect relatively thin materials to metal bases, when access is available to both sides of the joint. Locking nuts in a variety of forms are available to counteract the possibility of loosening, caused by vibration. There are also special nylon-lined nuts and screws with nylon inserts, which achieve the same locking effect.

Note: Although machine screws are generally used in conjunction with a nut, they can sometimes be used as blind fixings into a threaded hole in the metal base. Where a threaded hole is of a poor quality, or a higher standard of mating between the thread and the tapping is required, a thread insert may be used in the tapped hole. This reduces the size of the machine screw, but can accommodate variations in pitch and diameter.

A3 Set screws

These are in effect headless machine screws and are used for the same general applications as headed machine screws. Usually the smallest diameters of set screw are more common, although the range of available diameters extends from M3 to M24. They can have slotted or hexagonal socket-driving profiles and square, cone or dog points. Set screws tend more frequently to make blind connections into tapped drillings in the metal base, than to be used with a matching nut.

Section B: Light fixings

This group of fixings covers devices which make a connection between a thin component (often a sheet of metal or other material, e.g. industrial claddings) and a heavier metal base. It also includes devices which make a connection between two thin metal elements.

In spite of the name of the group, the engineering performance of many of these devices should not be underrated. This is particularly true of some of the extensive range of devices used to attach cladding to steel framework. These have not only to resist considerable uplift forces caused by wind pressure, they often, too, are required to fix the cladding with sufficient firmness to enlist the strength of the cladding to brace the structural frame.

Most fixings in this group are base-deforming type 2(b) devices –

thread-forming, thread-cutting, self-drilling screws or powder- or pneumatic-actuated fixings. There is also a small group of type 2(a) drilled-for fixings which are associated with making single-sided, blind connections between a thin metal base and a thin component. These include light rivets, rivet bushes and self-locking tapped holes.

BI Base-deforming devices (type 2(b))

Fixing devices in this category include: self-tapping screws; self-drilling screws; powder-actuated fasteners; pneumatic-actuated fasteners.

Self-tapping screws

These screws all require a pilot hole of the correct diameter to be drilled in the base into which the screw is driven. During driving, the screw produces its own mating thread in the base material. Clearly, self-tapping screws are only effective if there is sufficient thickness of base material in which to form a thread long enough to give a firm fixing (Fig. 5.12).

Description Self-tapping screws are usually made of hardened or stainless steel and fall into two categories: thread-forming and thread-cutting screws. The latter type are not greatly

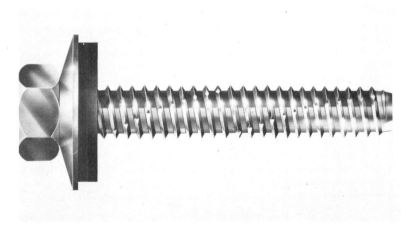

Fig. 5.12   Buildex ST and self tapping fasteners for roofing and cladding fixings

used in building. They form their mating thread by cutting and there are various point profiles, usually slotted or chipped, to aid the cutting action and to avoid the resulting metal swarf from clogging the screw.

Thread-forming screws, on the other hand, form their mating thread by deforming the base and not by removing any of its material. The two most commonly-used types of thread-forming screws are the type AB, with a widely-spaced thread and a gimlet point, and the type B (formerly Z) with a similar thread and a slightly tapered point. Head styles include countersunk, raised countersunk, mushroom, flange, pan and hexagonal; driving profiles are slotted or recessed. Sizes range from self-tapping screw gauge no. 2 (2.18 mm) to 14 (6.15 mm) and lengths from 5 to 50 mm.

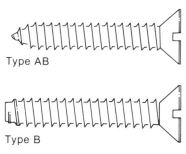

Type AB

Type B

Fig. 5.13   Self-tapping screws, types AB and B

Applications The type AB thread-forming screw is used in light sheet-metal bases; the type B in light and heavy sheet-metal and non-ferrous castings.

Setting instructions

1. Drill a pilot hole in the base material of the diameter recommended for the particular screw diameter. A clearance hole should be drilled in the subsidiary component.
2. Position the component and place the screw point in the pilot hole and drive home using a manual or electric screwdriver.

Note: A unique type of thread-forming screw is produced by GKN. Called the *Taptite*, this device, made of cold-forging steel, has a trilobular shank (in section rather like an equilateral triangle with its corners rounded off). This, when driven into a circular pilot hole, forms its mating thread by the action of the short forward face of each lobe (Fig. 5.14). This gives a high degree of thread engagement, reduced friction during driving and high clamping action of the subsidiary component to the base. It is also claimed that these screws have a tensile strength greater than a conventional self-tapping screw or machine screw.

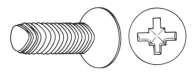

Fig. 5.14 Taptite self-tapping screws

The Taptite screw is available with various head styles and driving profiles, in diameters from M3 to M8 and lengths from 5 to 40 mm.

It can be used in thick steel sections, cored die-castings and thin sheet steel, provided that the pilot hole has an extruded rim. In the latter case a pilot hole is punched in the sheet steel so as to leave a protruding rim around the hole on one side of the sheet (Fig. 5.15). This clearly increases the 'thickness' of the sheet to receive the mating thread. Taptite screws can normally be used successfully with a drilled or clean (non-rimmed) punched hole when the thickness of the sheet is greater than three-quarters of the diameter of the fastener.

Another special type of thread-forming screw driven into a pilot hole in a metal base by a hammer, rather than a screwdriver, is the 'U' type of drive fastener. This is referred to in BS 4174:

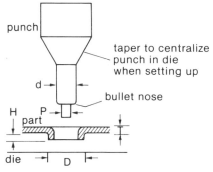

Fig. 5.15 Extruded hole to receive a Taptite screw

1972 as a 'metallic drive screw' and it usually has a round dome head. It is available in a range of diameters from self-tapping screw gauge no. 00 (1.5 mm) to 14 (6.15 mm) and lengths from 3 to 19 mm. This fastener, unlike the other self-tapping devices, is almost impossible to remove, once placed.

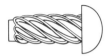

Fig. 5.16   U-type drive fastener

Self-drilling screws

Self-drilling screws are self-tapping screws which eliminate the need to drill a pilot hole. In fact three operations (drilling, tapping and placing) are included in one speedy fixing operation which results in high-performance connections which are particularly appropriate for firm cladding fixings. Self-drilling screws consist of a drill section whose length varies according to the thickness of steel base to which the fixing is to be attached. The drill section contains a drill flute, longer than the thickness of the steel to be drilled, and which provides top rake for the efficient cutting action and an exhaust passage for the drilling swarf (Fig. 5.17). Above the drilling section is the threaded part of the shank which taps its own mating groove in the base. Because the screw threads advance the fastener approximately ten times faster than the drill point can remove the metal, drilling must be completed before the

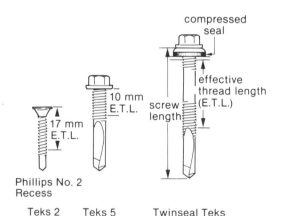

Fig. 5.17   Self-drilling screws (Twinseal, Teks 5 and Teks 2)

threaded section commences forming its mating thread. This makes the choice of the correct screw, which combines the necessary drilling and threading lengths, essential. Also the threaded portion of the shank has to penetrate the full depth of the drilling to realize the full holding power of the fixing.

Self-drilling screws are usually placed using an electric screwdriver.

Description Teks self-drilling screws are made of medium carbon cold forming steel, and have been heat treated. They are supplied in a variety of permutations of drilling and threaded section lengths to suit various thicknesses of steel base and component. They also have head/washer styles to match waterproofness need. The Sela range of British Screw and the Teks range from Buildex are the two most commonly-used families of self-drilling screws.

Twinseal cladding screws (Buildex) are available in two general ranges of self-tapping screw gauge no. 12 and lengths from 25 to 100 mm. There are some other gauges available for special purposes. The Twinseal screw has a metal/EPDM hexagonal head and sealing washer. Another Teks Screw – the Scots Teks – has a stainless steel head and is available in 25, 32 and 38 mm lengths. In addition there are similar ranges of deck-fixing self-drilling screws, including hexagonal heads or round heads with recessed driving profiles.

Applications Teks self-drilling screws are used primarily for fixing cladding or decking to steel framework in one operation, without prior preparation, and from the outside of the cladding or the decking. One range of Twinseal cladding screws is intended for fixing to cold-rolled purlins up to 3.5 mm thick; another to hot-rolled purlins from 5 to 12.5 mm thick. There is also a range of Standoff Teks intended to fix cladding and insulation in one operation without compressing the insulation (Fig. 5.18).

It is important to use the correct power screwdriver when fixing self-drilling screws. Buildex and British Screw recommend the Black and Decker HV25T because it has been designed to provide correct speed, power and a depth-sensing clutch to facilitate correct driving. A full range of accessories for use with this screwdriver and to suit both manufacturers' screws is available.

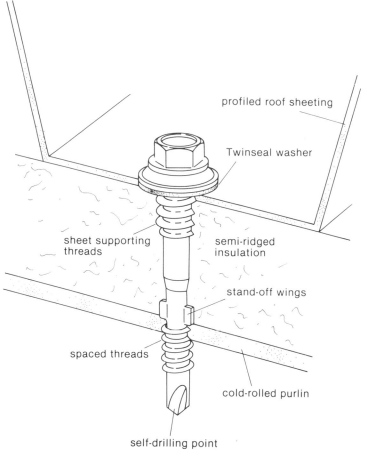

profiled roof sheeting

Twinseal washer

sheet supporting threads

semi-ridged insulation

stand-off wings

spaced threads

cold-rolled purlin

self-drilling point

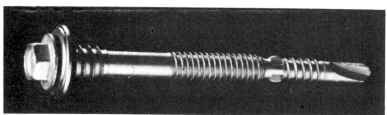

Fig. 5.18 The Standoff Teks

Setting instructions

1. It is particularly important when fixing cladding fasteners that consistent pressure on the sealing washer is main-

tained. In order to achieve this, always use an electric screwdriver with a depth-locating nose.

2. Adjust the depth-locating nose in accordance with the manufacturer's instructions for the particular device (Fig. 5.19).

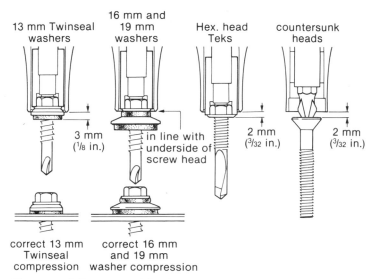

Fig. 5.19 Adjustment of depth-sensing nose on electric screwdriver

3. Holding the screwdriver perpendicular to the component, drill through the cladding and metal base until the screwdriver's nose makes contact with the component. The drive will then automatically disengage.

4. Check that the washer under the screw head is firmly bedding down. If not, make any necessary adjustment to the depth-locating nose.

Note: There is a number of special washers available for use with all self-drilling screws, either for weathersealing or spreading the fixing load over soft insulation boards (Fig. 5.20). There is also a number of snap-on colour caps for use with cladding fasteners.

Powder-actuated fasteners

These fixings require no prior preparation of the subsidiary com-

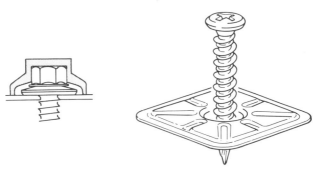

Fig. 5.20   Dec-Loc washer with Hilo self-drilling screw and
British Screw head protective cover

ponent or the base, such as the drilling of pilot holes, as in the case
of self-tapping screws. The fixing is merely driven through the
component and into the base (or merely into the base) by an
explosive charge from a cartridge in the fixing tool or gun. Fixings
take two forms:

1. Galvanized steel drive pins for direct firing through the
   subsidiary component to create a permanent fixing to the
   base in one operation.
2. Galvanized steel threaded studs for firing into the base to
   give firm fixing points to which to make detachable
   fixings to the subsidiary component. There are accessories, such as ring or eye couplers, clamps and hooks
   which can be used with threaded studs.

Both types of fixing (Fig. 5.21) achieve their holding power by
the plastic deformation of the steel, whose elasticity causes the
shank of the pin to be gripped immediately after it is driven.
   Typical subsidiary components which can be fired through to
make a direct, one-operation fixing to a metal base include:

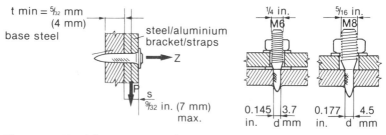

Fig. 5.21   Fired fixing types (a) drive pin (b) threaded studs

(a) wood (softwood or hardwood);
(b) metal sheeting or brackets;
(c) soft sheet materials (insulation board, wood wool, cork, etc.) usually employing an additional washer to avoid punching through.

Different sizes and shapes of drive nail are produced to match the character and thickness of the subsidiary component material.

Further information on the use of powder-actuated fixing in metal bases, their positioning and other matters affecting their performance are dealt with in the background section of this chapter on page 141.

Description Powder-actuated fixings are drive pins or nails of hardened (austempered) steel, fired from a tool by an explosive cartridge to make a permanent fixing into a metal base.

Pins are produced with various head and shank styles to suit the material with which they are used. Shank diameters are 3.7 or 4.5 mm and lengths are obtainable up to 65 mm. Plain-headed pins are intended to fix the subsidiary component direct to the base. Threaded stud-headed pins are available in stud diameters from M4 to M8. These are intended to provide a firm fixing to which to make removable connections for the subsidiary component. There are various colour-coded strengths of cartridge available to give the necessary force to drive the length of pin needed through component and base satisfactorily.

Applications Powder-actuated pins can make medium-duty fixings (limited to 15 kN pull-out strength) to metal bases over 4 mm thick, provided the metal in question is not brittle (e.g. cast iron) and has an ultimate tensile strength greater than 10 kgf/mm^2 (kilogram force per mm^2). The strength of the fixing will be dictated by the thickness and the strength of the metal base and the resistance of the subsidiary component to pull-through (in the case of direct fixings), or the strength of the nut connection to the threaded stud in the case of a removable fixing.

To obtain the most effective pull-out resistance, the point of the pin should pass right through the steel base, otherwise the compressive forces in the steel, acting on the point area, will tend to force the pin back out. When fixing to very thick steel bases, this proviso may not be valid, provided that the depth of penetration of the pin is sufficient to overcome the negative force on the embedment point. It is important that the diameter of the shank of the pin should not exceed the thickness of the steel base to obtain the best holding power.

If fixings are to be driven into metals other than structural steel, it is wise to carry out a suitability test on hardness, and if in doubt, consult the supplier of the metal.

The success of these fixings also depends on the type of fixing tool and the positioning of the pins and the thickness and type of subsidiary component.

Fixing tools

These have been described in some detail in Chapter 3, page 47, together with advice on their use. These pages should be referred to before attempting to make powder-actuated fixings.

It should be remembered that the Powder Actuated System Association (PASA) produces basic safety information and training data. Operating instructions should be followed meticulously and only fasteners and cartridges compatible with the tool should be used. Also only skilled operatives should be in charge of powder-actuated tools.

Positioning of pins

The minimum distance between the pin and the edge of the base should be two-and-a-half times the diameter of the shank. The minimum spacing of pins should be six times the diameter of the shank.

Thickness and type of subsidiary component

When the component is fixed direct to the mass walling, its thickness and type is critical to the success of the fixing. If the material is too thin, the pin may punch through (as in the case of very thin metal) or splinter (as in the case of wood). If the material is too thick, the fastener could bend, or not penetrate the base sufficiently, even using a pin with the longest shank.

Manufacturers' data sheets give full details of the thickness of subsidiary components which can be fixed using various lengths of fastener.

When soft materials are being fixed, such as insulation board, with the risk of pull-over failure, or firing through, additional washers may be required (Fig. 5.22).

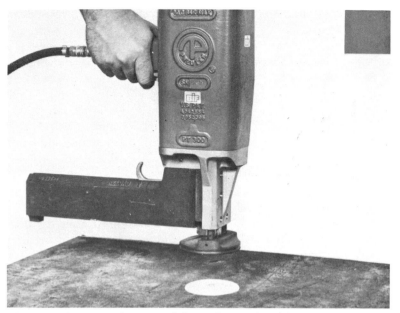

Fig. 5.22 A pneumatic-actuated fixing for insulation board

Setting instructions
Always follow the manufacturer's operating instructions minutely.

1. Select the length and type of pin required using the manufacturer's data, bearing in mind the type and thickness of subsidiary component.

2. To ensure the correct cartridge is used to give the right penetration depth, commence with a test-firing, using a low-strength cartridge. If the pin does not penetrate so that its point passes through the base (or penetrate the base sufficiently, if the base is thick), change to a higher strength of cartridge. The various strengths of cartridge are colour coded.

3. Place the tool firmly against the base (or, if firing through the subsidiary component, the component), ensure the tool is at right angles to the base, and fire.

Note: It is wise to wear safety glasses while making powder-actuated fixings and, in enclosed surroundings, ear protection too. Do not attempt to drive into areas of metal that have been welded or torch-cut – the material here may prove too hard. Do not fix, to a steel base, material which is thinner than the diameter of the shank of the pin. Do not use pins with a shank longer than is required.

Further safety advice is contained in the Health and Safety Executive Guidance Note PM 14, *Safety in the Use of Cartridge Operated Fixing Tools*, the Powder Actuated Systems Association's *Guide to Basic Training* and *Site Safety* by Jim Laney in Longman's Site Practice series.

Pneumatic-actuated fasteners

These fixings are identical to powder-actuated fasteners; the only difference is the method of driving the pin. All parts of the preceding section can, therefore, be considered as applicable to pneumatic-actuated fixings, with the exception of that part dealing with the firing tools (Fig. 5.23). Details of these are included in Chapter 3, page 50.

Sheeting fasteners

As a major application of the previous group of fixings is to connect cladding (roof and wall) to steel framework, it is appropriate to mention a group of fasteners here which fall into neither base-deforming nor drilled-for categories.

These fasteners were developed largely for the fixing of corrugated asbestos-cement sheeting and are all based on a simple hook bolt which hooks round one leg of the sheeting rails or purlins. The bolt is threaded on its non-hooked end and this end passes through a drilling in the sheeting to receive a washer and nut, which holds the sheeting in place. The waterproofing of the fixing is then completed by a plastic cap snapping ovet the nut (Fig. 5.24).

B2 Drilled-for fixings

These fasteners are mainly used to secure two thin sheets of metal where a self-tapping screw would not be able to make a sufficient length of mating thread in the base to obtain a firm anchorage, or where the use of a machine screw and nut is prohibited because

Fig. 5.23 BIF pneumatic fixing tool at work

the manipulation of the fasteners is restricted to one side of the base.

Many fasteners to light-gauge metal bases are of the rivet type, most of which are not greatly used on site in the building industry. Some other devices need sophisticated equipment, working from both sides of the base sheet, to prepare the base for a later fixing. An example of this is a rivet bush which is forced into a pre-drilled hole and is clinched there by a press (Fig. 5.25). It forms a reusable threaded bush in the metal base that can receive repeated machine screw fixings. Usually, though, these devices

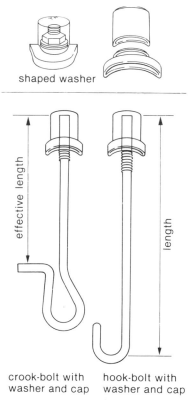

shaped washer

effective length

length

crook-bolt with
washer and cap

hook-bolt with
washer and cap

Fig. 5.24 Sheeting hook bolt and waterproofing cap

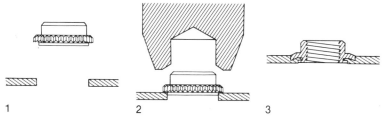

1   2   3

Fig. 5.25 Setting diagram for a rivet bush (ISC Press Nut)

are installed in the base sheet before it is delivered to site and therefore these are not applicable to this book.

Other devices which would be suitable for making connections to thin metal bases depend for their effect on the bunching of a flexible part of the device behind the base. These are in many respects indistinguishable from some of the devices discussed in Chapter 6. In making a selection of this type of fixing, therefore, it is advisable to examine the relevant parts of Chapter 6 as well as this section. Only those fasteners specifically developed to make 'blind' connections to thin metal bases (tapped holes and blind rivets) are considered here.

Self-locking tapped holes

These devices have a similar purpose to that of rivet bushes in that they provide a permanent, reusable, threaded fixing point to receive a machine screw. Unlike rivet bushes, however, self-locking tapped holes can be placed from one side of the base and without using a press. They fall into two basic types: screw actuated and independently placed tapped holes.

Screw-actuated tapped holes

Description There are various proprietary forms of these fasteners. Generally they consist of a brass, aluminium or plastic internally threaded cylinder with a flange at one end. The cylinder is shaped externally and split to encourage the opening of the fixing at the rear of the base when the screw is inserted.

Applications These fasteners are used to provide tapped holes in thin metal bases from 0.5 to 3.0 mm thick to receive machine screws of about M5 diameter. Some plastic tapped holes are used with 3.10 mm woodscrews (Fig. 5.26).

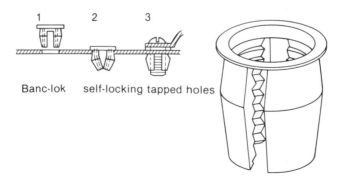

Fig. 5.26 Screw-actuated tapped hole

1. In the base, drill a hole of the diameter specified by the fixing manufacturer.
2. Insert the fastener so that its flange rests on the surface of the base.
3. Pass the machine screw through the component and screw into the tapped hole until tight.

Independently placed tapped holes

Description These devices are usually made of alloy steel or stainless steel and consist of an internally threaded tube with a profiled flange on one end. This flange appears on the base as being a flat or countersunk head.

Applications These fasteners make a fixed, anti-rotational tapped hole in this bases which will receive machine screws from M2.5 to M16. They are usually placed using a special handtool (Fig. 5.27).

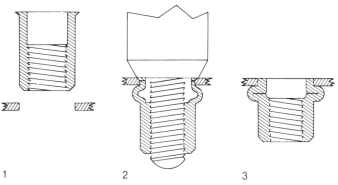

1   2   3

Fig. 5.27 Independently placed tapped hole (Clufix; ISC)

Setting instructions

1. In the base, drill a hole of the diameter specified by the fixing manufacturer.
2. Insert the fastener so that its flange rests on the surface of the base.
3. Using the handtool provided, drive the fastener into the base. This forces the profile under the flange into the base

material and avoids its rotation when the fastener receives the machine screw. It also bunches the fastener against the rear of the base.

4. Pass the machine screw through the component and screw into the fastener.

Note: A similar device sets a threaded stud in sheet metal, 0.5 mm and more thick, by buckling its shank at the rear of the base with a handtool (Fig. 5.28).

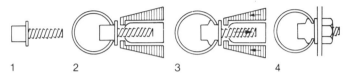

Fig. 5.28 Independently placed threaded stud (Rifbolt)

Blind rivets

Description All rivets depend on the enlargement of a part of the rivet on the further side of the base from the insertion side. There is a variety of lightweight rivets used in the manufacture of many factory-made products. Few of these are normally used on site. Those that are, are usually made of a variety of soft metals or plastics and either consist of a plastic insert which is hammer-set (Fig. 5.29) or a metal pop rivet which is bunched behind the base by the withdrawal of an integral steel pin using a special hand machine (Fig. 5.30).

Applications Most light rivets are used to connect thin sheets of metal or plastic of a combined thickness from 1 to 8.5 mm (hammer-set variety) or up to 12.5 mm (pop rivet variety).

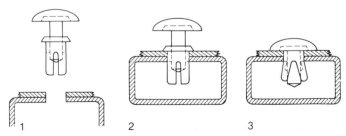

Fig. 5.29 Blind rivet (hammer-set variety) (Clic rivet)

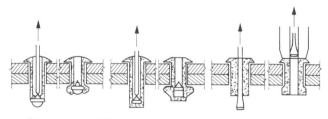

(a) Pop rivet   (b)Sealed-end rivet   (c) Automatic-feed rivet

Fig. 5.30   Blind rivet (pop rivet type)

Setting instructions

1. Drill a hole of the diameter specified by the manufacturer through both component and base.
2. Align holes and insert the rivet until its shoulder lies on the surface of the component.
3. In the case of hammer-set rivets, drive the head of the plunger until it is flush with the surface of the component; in the case of pop rivets, draw the steel pin upwards, using the rivetting tool, until the pin breaks.

Note: Pop rivets come in a variety of types, including ones with a sealed end (which are air- and watertight) and the automatic feed rivet. This latter is set by the withdrawal of a setting mandrel which is retracted, unbroken, into the fixing machine, unlike the normal pop rivet in which the pin is integral with the fixing and breaks-off on setting.

Self-tapping inserts

There is one further variety of device which is similar to a self-locking tapped hole, but which is used to form a threaded socket in thicker soft-metal bases. It is called the self-tapping insert.

Description These cylindrical devices are made of brass or case-hardened mild steel. They are threaded on the outside with a self-tapping thread, and on the inside with a thread to receive a machine screw. The inside thread is also cut by a series of vertical grooves which receive the edges of a hexagonal driving piece used to drive the socket into the base.

Applications The mild steel version of this device is used to create a tapped hole in a light alloy, the brass version in a plastic base. They vary in length from 9.6 to 15.7 mm and are intended to be used in bases which are thicker than these lengths. Self-tapping inserts are supplied in a range of sizes to receive machine screws from M4 to M10 (Fig. 5.31).

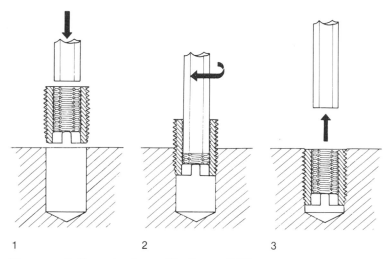

1   2   3

Fig. 5.31   Self-tapping insert (Kwik-sert; ISC)

Setting instructions

1. Drill a hole in the base of the specified diameter. The length of the drilling should be 1.5 mm longer than the insert.
2. Screw the self-tapping insert into the hole using a hexagonal wrench or an adaptor on a mechanical screw driver
3. When the top of the insert is flush with the base, stop screwing and remove the drivepiece.

Section C: Background

Bolted connections

Heavy-duty bolted connections, being of major structural import-ance, are always described by the engineer in detail on drawings and specifications. These instructions should be followed scrupulously. The design of bolted connections of this type is beyond the scope of this book. It might, however, be useful to learn a little of the background to these fixings.

Heavy metal-to-metal fixings derive from the peg, used in ancient timber construction. This later developed into the rivet in metalwork – a peg whose end was enlarged by beating on either side of the pieces of metal being joined. Nevertheless rivetted joints took a considerable time to form and did present a fire hazard, as a furnace was always needed close to the point of installation. The bolt, therefore, was developed which had most of the characteristics of the rivet, but few of its hazards.

The rivet has now almost completely disappeared from present-day construction, except for very light versions of the form (such as pop rivets, etc.), and even these tend to be used more frequently in equipment manufacture rather than on the construction site.

A bolted connection's strength depends on the size, type and number of the bolts, the thickness and type of steel (or other metal) from which the base and component are made and the spacing of the bolts.

The latter aspect is particularly important as the majority of steel-to-steel bolted connections put the bolts into single or double shear and their failure can result not only from the fracture of the shank of the bolts, but (more likely) from the deformation, splitting or eventual tearing through the base or component of the bolt. This is known as bearing failure.

As a result, the disposition of the bolts in relation to the thickness of the thinnest outer plate of the connection (for this purpose referred to as the component) is vital, and the assessment of minimum and maximum end and edge distances and centre-to-centre spacing of bolt becomes a balance between bolt diameter and thickness of the component. This is all laid down in the relevant British Standards and is too complicated to consider in this book. It is also unnecessary, as all bolted connections should be carefully designed by the engineer and thoroughly described, so that the work on site can be undertaken adequately.

The important factors to remember are:

    (a) the bolt holes should be accurately drilled to the correct diameter;
    (b) the bolt should fill the hole with practically no tolerance;
    (c) bolt holes must not be misaligned;
    (d) the nuts should be fully tightened – if high-strength bolts

are used, all the relevant advice given earlier in this chapter should be minutely followed.

Note: The method of designating the strength grade of bolts is worth a brief note. Grade designations are laid down in two digit numbers. The first digit is one-tenth of the minimum-strength grade of the steel in kilogram force per mm² (kgf/mm²). The second digit is arrived at by the formula:

$$\frac{1}{10} \times \frac{\text{yield stress}}{\text{minimum tensile stress}} \%$$

For instance, strength grade 4.6 (the normal grade for black bolts) indicates that the steel has a tensile strength of 40 kgf/mm² (first digit = 1/10 × 40 = 4) and a yield stress of 24 kgf/mm² (second digit = 1/10 × 24/40 × 100/1 = 6).

Strength grades of nuts are designated by one digit which represents one-tenth of the proof stress in kgf/mm². This corresponds with the minimum tensile strength of the highest grade of bolt with which the nut can be used.

The use of the kgf/mm² unit is in accordance with continental practice, but it is expected that the correct SI unit (N/mm²) will eventually be substituted (1 kg = 9.81 N).

Powder- or pneumatic-actuated fixings

There is a number of factors which affect the performance of fired fixings in metal bases. These have been referred to briefly and in a general way in the preceding section of this chapter. For those who require more detailed background information, this section has been included.

The following factors will influence the fastener's performance:

(a) strength and thickness of the metal base;
(b) the brittleness of the base;
(c) its yield point;
(d) the temperature at which the fixing is to be loaded;
(e) fastener spacing and positioning.

Strength and thickness of the metal base
Generally, fired fixings can be made to metal bases of 4 mm thickness or above, but the tensile strength of the metal has a great influence on the strength of the fixing and, generally, the lower

the ultimate tensile strength of the metal, the greater the thickness of material needed to make a satisfactory fixing. Fired fixings should not be attempted to metals with an ultimate tensile strength less than 10 kgf/mm².

With regard to firing into steel, manufacturers produce data sheets which relate the grade of steel to the thickness to which a fastener can be driven. These establish application limits that should be observed. Particular attention should be paid to these when fastening to steels with an ultimate tensile strength higher than 37 kgf/mm².

If a fired fixing is attempted to steel less than 4 mm thick (or to greater thickness of other metals of a lower strength) holding power will be diminished and denting, or bulging of the base material, may result. Also unacceptable crushing or deformation of the steel may occur when side loads are applied. Using an over-powerful cartridge may also bring the risk of firing through when thin components and bases are being fixed.

The best pull-out loads are obtained when the point of the fastener just breaks the underside of the steel base. When the metal base has unknown strength and hardness characteristics, a suitability test based on trial fixings should be made.

Brittleness of the base

Fired fixings should not be attempted to brittle metals. Good fixings are difficult to obtain and accident hazard from nail breaking is considerably increased.

Yield point

Steel with a low yield point can be subject to deformation when fasteners are loaded laterally (in shear or in bending). In abnormal conditions, this can result in the loosening of the fastener.

Use temperature

There is a sudden and noticeable drop in the holding power of fired fixings in steel bases when the temperature exceeds 250 °C. Generally, use temperatures between minus 10 °C and + 250 °C. do not affect either the penetration of the fixing or its holding power. Below minus 10 °C the toughness/ductility of steel decreases and fired fixings should not be used.

Fastener spacing and positioning

The hole produced by a fired fixing will have little or no effect on the structural performance of the base. Maximum effect will be in the order of less than 10 per cent reduction in strength. What is more, the use of fired fixings produces a better stress distribution in the steel than when a nut-and-bolt assembly is used. This is because the base suffers minimal loss of cross-section and, at the point where the fastener is driven, there is an increase in strength of the steel due to what is, in effect, the steel being worked cold by the action of driving the pin.

To obtain the best results and to avoid loosening already placed fasteners, the following spacing and positioning dimensions should be observed:

(a) distance between fastener and edge of steel base: 25 × shank diameter;

(b) distance centre-to-centre of fasteners: 6 × shank diameter.

Galvanic corrosion

When the fixing is made of steel and the base is also steel, clearly no question of galvanic corrosion exists. However, if the metal base is not steel, or the subsidiary component is of another metal, this danger should be checked out. Brief details of galvanic or bi-metal corrosion are contained in Chapter 2, page 9.

6

Mechanical fixings to thin-walled cellular bases (and thin sheet materials)

With the increasing use in the last 20 years of structures consisting of light framework faced with thin sheet materials and the introduction of a number of cellular composite products, such as hollow core flush doors and cellular composite partition panels, the problems of fixing to thin sheet materials began to receive priority attention from fixings manufacturers.

Previously fixings to sheet materials had not been made to the facing material itself, but to its supporting framework. As a result, fixings had to be located where there were studs or noggings behind the sheeting, or additional framework had to be inserted in the assembly on purpose to receive the fixings. This restraint was becoming increasingly unacceptable. Means had to be discovered to fix directly to thin sheet facings, like plasterboard, plywood, chipboard, asbestos cement, metal and plastic sheets, without having to search for a framing member behind the sheeting. What is more, in the case of pre-formed composite materials with facings on both sides of a cellular core, there no longer were any substantial internal framing members to receive fixings; and most fixings had to be placed from one side only of the base. In other words, the fixings had to be 'blind'.

The result has been the development of a number of types of fixing – sometimes called cavity fixings – which can be used to make relatively light-duty, but reliable, fixings to thin materials or thin-walled composites.

All types are installed in pre-drilled holes (making them, according to our classification method, type 2a fixings) and all achieve their holding power by clamping the subsidiary component to the surface layer of the base and spreading its load as widely as possible over that layer. This is done by opening or expanding a part of the device against the reverse face of the sheet material. Some of these devices can only be used once, because a part of the fixing is lost in the cavity when the subsidiary component is unfixed. Others are wholly reusable.

Fasteners in this group fall broadly into four subdivisions:

(a) those which operate by the expansion of a flexible part of the device behind the facing sheet;
(b) those which depend similarly on the opening of a wedge-shaped obstruction behind the fixing hole, as in a metal rivet;
(c) those which involve the turning or opening of a toggle within the cavity;
(d) the so-called 'umbrella' cavity fixings, which, during setting, fan out a series of spokes or legs inside the cavity to anchor the fixing.

Table 6.1 lists the broad categories of cavity fixings and gives some guidance on which type should be used in which set of circumstances.

Section A: Expanding cavity fixings

This group of cavity fixings includes at least one – the Rawlnut – which can be used as a drilled-for fixing in a mass walling base. It achieves its holding power in a cavity context by the expansion of a captive rubber (or neoprene) sleeve behind the facing sheet; but in addition to this type of expanding cavity fixing there are those whose setting causes a strong, split nylon sleeve to bunch against the back of the facing sheet; others which are similar to plastic plugs and expand inside the cavity when a loose screw is inserted, and those which spread legs against the back of the surfacing material.

All these fixings have a reuse value, except for the last type.

A1 Expanding sleeve (Rawlnut)

Description This device consists of a tough rubber sleeve, into one end of which is bonded a metal nut. The other end of the sleeve is flanged to stop it falling through the drilling hole. A machine screw with an ISO metric coarse thread is inserted into the sleeve and, as it is screwed into the nut, the sleeve is expanded

Table 6.1 Cavity fixings: which fixing to use

Description of base	Expanding sleeve	Split sleeve	Plastic plug	Legged plug	Rivet	Gravity toggle	Spring toggle	Cord retention toggle	Plastic toggle	Umbrella fixing
Plasterboard	✓	✓	✓	✓		✓	✓	✓	✓	✓
Cellular partition	✓	✓	✓	✓		✓	✓	✓	✓	✓
Lath and plaster	✓		✓			✓	✓	✓	✓	✓
Insulation board	✓	✓	✓	✓		✓	✓	✓	✓	✓
Hardboard	✓	✓	✓	✓		✓	✓	✓	✓	✓
Asbestos/silicate board	✓	✓	✓	✓		✓	✓	✓	✓	✓
Plywood	✓	✓	✓	✓	✓	✓	✓	✓	✓	✓
Laminated materials	✓	✓	✓	✓		✓	✓	✓	✓	✓
Plastic sheet	✓	✓	✓	✓	✓	✓	✓	✓	✓	✓
Glass	✓	✓		✓					✓	
Sheet metal	✓	✓		✓	✓					
Hollow core doors		✓			✓					
Retrievable	✓				✓				✓	
Unretrievable		✓	✓	✓		✓	✓	✓		
Permanent fixing even when unscrewed	✓	✓	✓		✓					✓

against the rear surface of the lining material. Rawlnuts are available in diameters from 8 to 24 mm and with sleeve lengths from 11 to 27 mm. A neoprene sleeve version is also available for locations where oil, high ozone content or high temperatures could present a risk to a rubber sleeve.

Applications Rawlnuts make demountable fixings (i.e. the expanding sleeve can be left behind in the wall to receive further fixings, or the whole device can be removed and reused elsewhere) in all types of thin materials or cellular bases. They are also vibration resistant, electrically insulating, waterproof and corrosion resistant (Fig. 6.1).

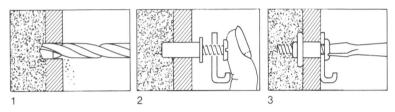

Fig. 6.1 Rawlnut setting diagrams

Setting instructions

1. Drill a hole of the diameter recommended by the manufacturer.
2. Push in the sleeve so that its flange rests against the surface of the base. Position the subsidiary component and pass the fixing screw through it and into the sleeve.
3. Tighten the screw.

Note: Rubber-sleeved Rawlnuts can be used with safety in areas where the temperature rises to 70 °C. They are not, however, oilproof. Neoprene-sleeved devices are unaffected by the air temperature up to 90 °C and are resistant to all types of oil up to temperatures of 110 °C.

A2 Split-sleeve cavity fixing (Thuscan Rosegrip)

Description This type of device, like the previous one, consists of a plastic sleeve with an integral nut cast into one end and a flange formed in the other. In this case, however, the sleeve has a series of longitudinal splits. A machine screw is inserted into the sleeve and, as it is screwed into the nut, the sleeve is drawn towards the surface material and the plastic strips are bunched behind the facing in the form of a rosette. These devices are available in a single screw size (5 mm diameter) and in 3 lengths, selected according to the thickness of the facing material – 3 to 8 mm; 8 to 16 mm and 16 to 26 mm.

Applications Split-sleeve devices make demountable fixings (i.e the sleeve remains in the wall and can receive further fixings, if required) in all types of thin material or cellular bases within the size limitations set out above (Fig. 6.2).

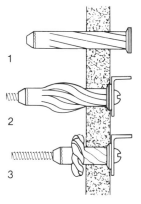

Fig. 6.2 Thuscan Rosegrip setting diagrams

Setting instructions

1. Drill a 10 mm diameter hole (or as recommended by the manufacturer). Push in the sleeve so that its flange rests against the surface of the base.
2. Position the subsidiary component and pass the fixing screw through it and into the sleeve.
3. Tighten the screw.

A3 Plastic plugs (Fischer NA Nylon rivet, Rawlanchor)

Description This is a nylon plug, internally threaded to accept a standard woodscrew. One end of the plug is flanged to avoid it falling into the cavity. Three sizes of woodscrew (3, 4 and

150

5 mm diameter) can be accommodated in a range of 7 plug sizes and 4 lengths. The length is selected according to the thickness of the facing sheet from 3 to 26 mm. As the woodscrew is driven into the plug, so the plug is drawn towards the facing material and folded against its rear surface.

Applications NA Nylon rivet anchors make demountable fixings (i.e. the expanded plug remains in the wall and can receive further fixings if required) in most types of thin material or cellular bases. The nylon is strong, durable and will not corrode (Fig. 6.3).

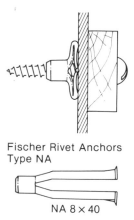

Fischer Rivet Anchors
Type NA

NA 8 × 40

Fig. 6.3 NA Nylon rivet (Fischer)

Setting instructions

1. Drill a hole of the diameter recommended by the manufacturer for the particular size of anchor.
2. Push in the sleeve so that its flange rests against the surface of the base.
3. Position the subsidiary component and pass the woodscrew through it into the plug.
4. Tighten the screw. Do not overtighten.

Note: The anchors have a moulded thread which accepts the woodscrew. It is important that the correct size of screw is used and that it is not over-tightened. Also the plain (unthreaded) length of the shank should not exceed the combined thicknesses of the subsidiary component and the facing layer.

A4 Legged plugs (Fischer Type A hollow door fixing)

Description This nylon device is internally threaded to accept woodscrews and is available in 4 sizes for use in small cavities with screw sizes from 3 to 5 mm diameters. It has a series of outward-facing legs which are drawn back against the inside surface of the facing material when the screw is tightened.

Applications This type of anchor was developed specially for use in narrow cavities (between 15 and 20 mm wide) – the sort of size normally encountered in hollow-core flush doors. When the screw is unscrewed, the nylon body falls inside the cavity, making this device non-reusable. The thickness of the facing layer, however, is not critical to this fixing, provided that the length of screw will accommodate the thickness (Fig. 6.4).

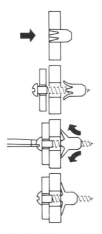

Fig. 6.4 Fischer Type A hollow door fixing

Setting instructions

1. Select a screw of the diameter appropriate to the nylon body and of a length equal to the thickness of the surface layer, plus the thickness of the subsidiary component, plus the length of the anchor.
2. Drill a hole of the recommended diameter.
3. Thread the screw through the subsidiary component and screw the anchor body on to the point of the screw until it is firmly gripped.

4. Position the subsidiary component, pushing the anchor through into the cavity.
5. Pull the screw outwards, drawing the anchor against the rear of the facing layer.
6. Tighten the screw.

Note: It is important that the recommended diameter of woodscrew is used.

Section B: Rivet cavity fixings

This group of fixings is designed specifically for connecting a thin component to a thin base, the combined thicknesses of the two being in some cases less than 4 mm. The action of rivet fixings is not dissimilar to that of the expanding cavity fixings above, the main difference being that the action of inserting a woodscrew or self-tapping screw into the device splays out the legs of the rivet in the cavity to form a wedge shape, which inhibits withdrawal.

Description Generally these devices are of nylon with a thin flange on their external end, the other end being split into a series of legs. These are splayed outwards by the insertion of a woodscrew or self-tapping screw (depending on the pattern of rivet). Sizes range from rivets to fit 4 to 8 mm diameter holes, and rivet lengths from 9.5 to 24.5 mm.

Applications These fixings are used to connect a thin (usually metal) subsidiary component to a thin base. Often both component and base are thin-gauge metal, in some cases the device can be used with thicker bases such as plywood or glass fibre sheet (Fig. 6.5).

Setting instructions

1. Drill a hole of the recommended size.
2. Insert the rivet so that its flange is set against the outer surface of the base.
3. Position the subsidiary component over the rivet and drive the screw into the device.

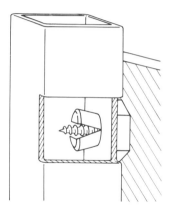

Fig. 6.5 Rivet cavity fixing

Section C: Toggle fixings

This group of mainly metal fixings includes those which open within the cavity automatically, either by the action of gravity or because of a spring in the device. They are suitable for most cellular materials provided the cavity is wide enough to allow the device to open. Some are not recommended for use with glass or metal bases. These fixings are generally not reusable, the toggle section remaining in the cavity if ever the assembly is dismantled.

C1 Gravity toggles

Description As the name suggests, these toggles open automatically as a result of the action of gravity. They consist of a plated steel toggle bar, pivoted off-centre on a swivel nut. This nut receives a machine screw, ranging in diameter from M3 to M6 and in lengths from 50 to 80 mm.

Applications Gravity toggles are applicable to most vertical cellular bases with soft facing materials like plasterboard, fibreboard, etc. provided that the width of the cavity will allow the bar to be inserted horizontally and then fall into a vertical position behind the facing sheet. The bar cannot be retrieved, if ever the fixing is disassembled, but will remain in the cavity (Fig. 6.6).

Setting instructions

1. Drill a hole in the base of sufficient diameter to take the

154

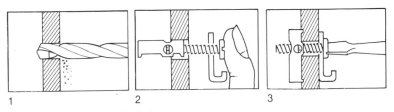

1   2   3

Fig. 6.6 Gravity toggle fixing

bar of the device.

2. Pass the machine screw through the subsidiary component and run the nut of the device on to the screw. Holding the bar horizontally, insert it through the fixing hole and allow it to fall freely within the cavity.

3. Draw the bar back against the inside face of the cavity and tighten the screw.

Note: This device will not operate when the base is lying in the horizontal plane. In that case a spring toggle should be used.

C2 Spring toggles

Description The spring toggle is often referred to as a 'butterfly' toggle and consists of a plated steel, hinged bar which is maintained in the open position by the operation of a spring. The toggle bar is pivoted on a swivel nut which receives a machine screw. Screw diameters range from M3 to M6 and lengths from 50 to 80 mm.

Applications This type is applicable to most vertical or horizontal cellular bases with soft facing materials like plasterboard, fibreboard etc, provided that the width of the cavity will allow the bar, which is folded for insertion, to open inside. The bar cannot be retrieved and will remain in the cavity whenever the machine screw is unscrewed (Fig. 6.7).

Setting instructions

1. Drill a hole in the base of sufficient diameter to take the bar in a folded condition.

2. Pass the machine screw through the subsidiary component and run the nut of the device on to the screw. Fold

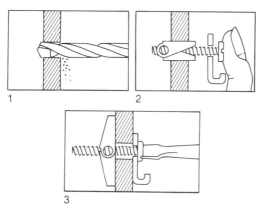

Fig. 6.7 Spring toggle

the hinged bar against the spring and insert it through the fixing hole until the spring is free to open the bar.

3. Draw the bar back against the inside face of the cavity and tighten the screw.

C3 Cord-retention toggles

Description Unlike the toggle fixings described above, this type of fixing has a notched nylon bar connected to a nylon cord running through a movable nylon guide sleeve. Toggle bars are either 31.8 or 57.2 mm long. This device makes a fixing point to receive a woodscrew.

Applications This type of device can be used in similar applications as the previous toggle fixings. It is particularly useful for making fixings in plasterboard ceilings. The fixing is non-retrievable (Fig. 6.8).

Setting instructions

1. Drill a hole in the base of sufficient diameter to take the toggle bar.
2. Manoeuvre the toggle bar into position using the attached nylon cord.
3. Slide the guide sleeve up the cord to close the fixing hole.
4. Pass the screw through the subsidiary component and

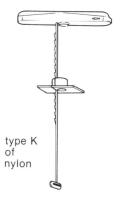

type K
of
nylon

Fig. 6.8 Cord-retention toggle

centre the screw accurately through the guide sleeve and into the threaded hole in the toggle bar.

5. Cut off the surplus length of nylon cord protruding from the guide sleeve before finally fixing the subsidiary component.

C4 Light plastic toggle plugs (Thunder toggle plug)

Description These are similar to the gravity toggle, but the toggle bar in this case is housed in the body of the plug and is made of plastic. A screw is supplied with the device and a plastic strap is attached to the head of the toggle bar. The body of the plug has a small flange to prevent the plug falling into the cavity.

Applications These toggle fasteners can be used in most cellular bases where the cavity width is at least 31 mm and the facing board is between 8 and 9.6 mm thick. Plastic toggles are retrievable (Fig. 6.9).

Setting instructions

1. Drill a hole of the recommended diameter. Insert the plug.
2. Push the screw into the plug forcing the toggle into a position parallel to the base surfacing layer. This will align the screw with a threaded hole in the toggle.
3. Thread the screw through the subsidiary component and insert the screw into the toggle bar.

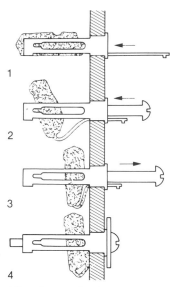

Fig. 6.9 Thunder toggle plug

4. Draw the toggle bar against the inside of the cavity using the screw and tighten the screw.

Note: Ensure that the plastic strap remains projecting from the plug during the whole of this operation and then conceal it behind the subsidiary component.

Removal instructions

1. Unscrew the subsidiary component.
2. Using the screw, push the toggle back into its original position parallel to the base surfacing layer and at the back of the plug body.
3. Using the integral strap, pull the bar back into alignment with the body of the plug.
4. Withdraw the plug.

Section D: Umbrella cavity fixings (Molly Nut, Rawlplug Interset, Fischer Fixoplac-metal anchor)

The fundamental difference between these fixings and expanding

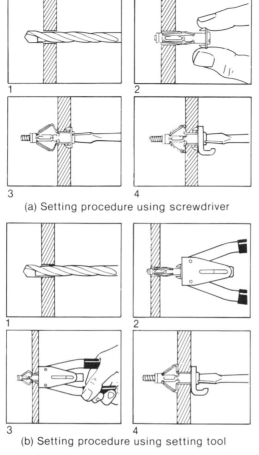

(a) Setting procedure using screwdriver

(b) Setting procedure using setting tool

Fig. 6.10 Umbrella cavity fixing (a) Setting procedure using
screwdriver (b) Setting procedure using setting tool.

cavity fixings is that the umbrella fixing provides a permanent threaded hole in the face of the base, supported within the cavity by a series of metal legs and from which a screw can be removed and replaced as frequently as is required. Both types of fixing do, however, have one characteristic in common; they both rely on the expansion of the device within the cavity.

Description These plated steel fixings consist of four folding legs attached at their near end to the flanged body of the device and at the further end to a nut which receives a machine screw of M4, M5 or M6 diameter. By screwing the machine screw into the nut, the legs are folded outwards and drawn back against the rear surface of the facing material. A setting tool is provided by some manufacturers to speed up setting operations.

Applications Umbrella cavity fixings can be used to make quick permanent fixings to plasterboard, plywood, chipboard and similar sheet materials (Fig. 6.10).

Setting instructions

1. Drill a hole of the diameter recommended by the manufacturer.
2. Insert the umbrella fixing.
3. Screw the machine screw into the device to fold out the legs. Alternatively a setting tool supplied by the fixing manufacturer can be used.
4. Remove the screw, thread it through the subsidiary component and screw it back again into the fixing to secure the component.

7

Adhesives

The range of available adhesives for use in building has considerably increased in recent years. Also the complexity of the chemical formulations involved in modern adhesives is so daunting that the subject can only be dealt with in general terms in a book of this type. It is sometimes difficult to allocate with certainty a proprietary adhesive to a general category, or even to be sure of its suitability for a particular use. All recommendations given here should be checked against the manufacturer's specific instructions for the adhesive concerned.

Some of the more sophisticated modern adhesives, such as some of the thermosets (adhesives whose chemical set is encouraged by heat), are difficult, or even impossible, to use on site. Most of them require careful mixing and application which could prove impractical in site conditions. A few give off highly flammable and toxic gases which can introduce problems of safety in their use and storage.

Nevertheless, adhesives have distinct advantages to offer when compared with mechanical fixings in certain applications, particularly where modern plastic materials are being fixed, or where visible fixings are unacceptable, or in cases where high-strength repairs are necessary to concrete structures. Adhesives also eliminate much of the preparatory work to making the fixing, such operations as drilling and plugging, merely requiring the surfaces which are to be joined to be scrupulously clean and possibly sanded and primed.

With the introduction of synthetic resin adhesives, reliable high-strength structural joints could be achieved with glue for the first time. This structural use of adhesives originated in the fabrication of timber components, such as laminated beams and stressed-skin decking panels which were mass-produced in factories. This use of adhesives to produce structural components has remained largely a factory, and not a site, practice. However, more recently there has been a marked increase in the use of adhesives on site in connection with structural concrete. We have already discussed one aspect of resin adhesives applied in this way when we considered chemical anchors in section A2 of Chapter 3. These structural adhesives are primarily intended for site use and will be dealt with later in this chapter.

Generally, though, adhesives tend to be used on site for making non-structural joints, often between the structure and an applied finish, like a flooring or a ceiling tile or a decorative plastic wall lining. Some adhesives are manufactured to fulfil one specific task; others may have a number of different applications.

As in mechanical fixings, there tends to be a heavy, immobile part of the joint – the base – and a lighter, more mobile component. Both can theoretically be referred to as *adherends*, but the heavier part is sometimes in adhesive jargon called the *substrate*. Generally, for consistency, we shall continue to refer to the substrate, where appropriate, as the base.

How to use adhesives

The most important piece of advice that can be given to anyone using adhesives is to follow the manufacturer's instructions. These have been arrived at not to make the use of adhesives unnecessarily complicated, but for very good chemical and physical reasons in order to achieve the most efficient bond between adherends.

The way an adhesive is used depends on the type of adhesive, whether it is a *contact adhesive*, a *wet-stick adhesive*, or one of the other application types. There are other ways of classifying glues – by their ingredients, or their method of setting – but these aspects can more usefully be examined in the background part of this chapter.

Contact adhesive

This type of adhesive, sometimes referred to as a *two-way dry-stick adhesive*, is used when both the base and the component are nonporous. Once a joint made with a contact adhesive has been closed, there is no chance of adjusting the position of one part of the joint relative to the other. In other words, contact adhesives have high grab. This is very useful for some types of work where clamping the parts of the joint together would be difficult; but it makes the correct positioning of the parts of the joint at the first attempt vital. Contact adhesives are usually neoprene rubber solutions in organic solvents and they give off flammable and toxic gases (see p. 172). Today a new breed of water-based contact adhesives is beginning to reach the market, but their use is limited to warm, indoor applications at present. Contact adhesives are used to fix plastic laminates, cork tiles, etc. in other applications where high grab is essential.

Method

1. Ensure the surfaces to be bonded are clean, free from dust, oil, grease or loose material.
2. Apply the adhesive to *both* surfaces to be joined in a thin, even coating to manufacturer's instructions. (This will often be done by means of a metal or plastic comb or notched trowel with notches of a specified size.)
3. Allow the adhesive to dry (its solvent to evaporate) for the period recommended by the manufacturer. This 'open time' is important: too short an open time will mean 'wet' adhesive being trapped in the joint, resulting in a poor bond and 'springing' of the component; too long an open time will mean that the two layers of dried adhesive will not fuse when brought together.
4. Bring the parts of the joint together and apply an even pressure over the whole area of the joint. It is important when sticking down large, thin sheet materials to place the sheet down starting from one side and working steadily across the length of the sheet, making sure no air bubbles are trapped in the joint.

Wet-stick adhesive

This adhesive, sometimes referred to as a *one-way wet adhesive*, is used when either the base or the component, or both of them, are porous, thus allowing the solvent in the adhesive to soak into the material. These adhesives have varying degrees of *grab* and usually allow minor adjustment to be made to the parts of the joint after closing. They also have varying degrees of 'thickness', the selection of this quality depending on the roughness of the adherends and whether the adhesive needs to fill gaps at the interface of the two materials (if the adhesive needs to be *gap-filling*).

Method

1. Ensure the surfaces to be bonded are clean, free from dust, oil, grease or loose material.
2. Apply adhesive to *one* surface to be joined in an even layer and to the thickness recommended by the manufacturer. (This will often be done by means of a metal or plastic comb or notched trowel with notches of a specified size.)
3. On highly porous surfaces it may be necessary to apply two coats of adhesive, the first being used to prime the surface.
4. Once more the manufacturer's instructions concerning open time should be followed. Emulsion adhesives on non-porous bases should be allowed to become tacky before closing; while epoxide and polyester-based adhesives should be closed immediately.
5. Apply an even pressure over the whole area of the joint after closing until the adhesive has hardened. This may involve the use of a roller. Depending on the degree of grab of the adhesive, position of the joint, weight of the component and the adhesive drying time, the joint may have to be clamped in position for some time.

Note: Gum spirit adhesives (used mainly for sticking-down linoleum) and some latex-based adhesives form surface skins after application. These must be broken on closing the joint, by sliding one surface against the other when placing the adherends together. If this is not done a poor bond could result.

Gun-grade adhesive

Unlike the two previous types of adhesive, this is not applied as a continuous coating, but in beads or blobs. It is used mainly to stick board materials, such as plasterboard or chipboard, to solid backings and has moderate gap-filling qualities. It is often neoprene rubber-based and has good wet-grab characteristics. This type of adhesive is often used for fixing skirtings and architraves, stair nosings, or refixing loose wood block flooring.

Method

1. Ensure the surfaces to be bonded are clean, free from dust, oil, grease or loose material.
2. Apply adhesive direct from the cartridge in a series of blobs to the back of the component and at spacing recommended by the manufacturer for the type of component. In some cases the recommendation may be for a continuous bead of adhesive round the edges of the component, with intermediate beads across the component. The bond develops more quickly in this case.
3. Slide the component into position, placing it about 25mm away from its final position and pushing it home. This movement breaks the surface skin of the adhesive. The manufacturer's guidance on open time should be observed.
4. Press into place with an even, overall pressure.

Note: Provided the component is not warped, bond will be sufficient to hold most components in place and work can proceed to the next component. If the component, however, is warped and there is a danger of it springing from the base, it will be necessary to pin it into position in the warped area, or the following procedure would be followed.

Apply the adhesive as above. Offer the component up to its final position so that the adhesive blobs or beads wet the base. Pull away and allow the solvent to evaporate for about 20 minutes. Then replace the component, at which time the grab should be sufficient to retain the component.

Foam pads

These are adhesive-impregnated foam pads used to provide an alternative to blob-applied gun-grade adhesive.

Method

1. Ensure the surfaces to be bonded are clean, free from dust, oil, grease or loose material.
2. Bond pieces of foam, at the intervals recommended by the manufacturer depending on the weight of the component to be fixed, to the base, using contact adhesive.

3. Apply a coat of contact adhesive to the outer face of the foam pads.
4. Place the component to be bonded briefly in position against the pads in order to transfer some of the adhesive to the component.
5. Apply further adhesive, if necessary, to the wetted areas on the component and also on the pads.
6. When tacky dry (or at a time recommended by the manufacturer), replace the board and apply pressure.

Note: This method can be used, when the surface of the base is uneven, by doubling-up the thickness of the pads in low-lying areas. A double-sided self-adhesive tape is also available as an alternative to the method given above. In this case merely peel off the protective paper from the adhesive surface before placing. The method is easy to use, but should only be applied to smooth surfaces and when the component to be fixed is reasonably light.

Adhesives used in structural repairs

Some adhesives can be added to sand/cement mixes to give a mortar of enhanced adhesion and one which can be used to effect thin repairs to elements like concrete structures and floor surfaces without the later danger of the repair cracking or curling. Polyvinyl acetate (PVA), epoxy resin, styrene butadiene (SBR) latex and styrene acrylic emulsion adhesives can be used in this context, but the former must not be used externally or in damp conditions.

There are many pre-mixed proprietary concrete repair formulations on the market. In all cases the manufacturer's instructions will need to be followed carefully and the surface to be treated should be clean and free from all loose material. Also, if reinforcement has been exposed, all rust will need to be cleaned off and the steel treated with a suitable rust inhibitor prior to the repair being carried out.

Adhesives as bonding agents

Adhesives are often used to form a bond between thin sand/cement screeds, cement renderings and gypsum plaster coatings and smooth, dense backing surfaces. This is particularly necessary when a screed is laid on a concrete subfloor which is no longer 'green' to avoid the screed cracking and lifting. PVA emulsion adhesives are often used as bonding aids, applied as follows.

Method

1. Ensure the surface to be treated is clean, free from dust, loose material and contamination.
2. Prime the surface with a little of the polyvinyl acetate adhesive diluted with water. Allow to dry.
3. Apply a concentrated solution of polyvinyl acetate adhesive to the surface using a brush or spray.
4. Trowel-on the rendering, screed or plaster. In the case of most popular brands of polyvinyl acetate adhesive, the finish should be applied onto the wet and tacky coat. A few manufacturers, however, specify that their products must be allowed to dry before applying the rendering, screed or plaster.

Note: In more extreme cases where a bond is considered difficult to achieve a key can be provided by mixing a slurry of polyvinyl acetate bonding aid and ordinary Portland cement. This is first brushed onto the base, before the finish is trowelled-on, either before or after the slurry has fully set, depending on the recommendations of the manufacturer. Polyvinyl acetate should not be used in wet locations. Here epoxy, styrene butadiene latex or styrene acrylic alternatives should be used.

Hot-melt adhesives

There is a growing use of hot-melt adhesives (ones which are applied hot and set on cooling) on site. This is the result of the introduction of a number of light, highly mobile electric-powered pistol applicators, like the BIF Hipermatic range (Fig. 7.1).

This technique was originally developed for factory use, but now is becoming more familiar on site, provided electric power is available. Pistols have various interchangeable application nozzles and BIF, as an example, supplies a range of eight different hot-melt adhesives for use with different adherends, such as wood, plastics, glass and metal. The adhesive is supplied in a solid cartridge, 44.5 mm diameter by 50 mm in length. This is inserted into the applicator and, after an initial warm-up period of about 3 minutes, is extruded on the pressing of the applicator's trigger.

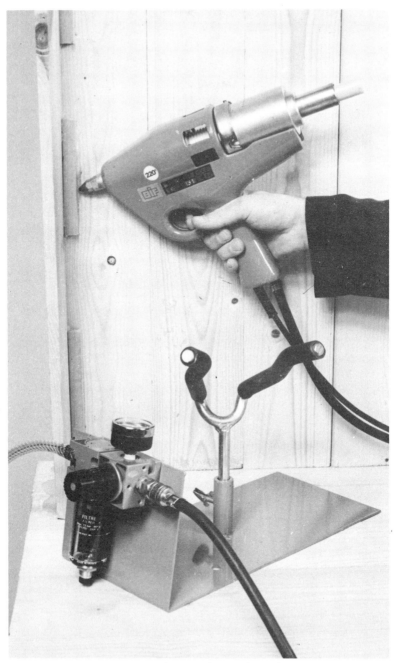

Fig. 7.1 Pneumatic BIF–Bond T 718 hot-melt applicator

Hot-melt adhesives have the benefit of being solvent-free, making their use and storage more safe, simple and practically odourless. It is also a less wasteful method of using adhesive, all the material being capable of being precisely placed where it is needed and without loss on tools, or in cans or mixing kettles.

General advice

In all cases where adhesives are to be used on site there are a few rules which should be observed:

1. Ensure that the adhesive you use is recommended for that particular application by the manufacturer. Is it compatible with the materials to be adhered to each other? Will it allow the necessary amount of adjustment to be made to the joint after it is closed? Has it the necessary gap-filling properties to suit the roughness of the adherends?
2. Always precisely follow the adhesive manufacturer's instructions in the use and application of the product.
3. All surfaces to be bonded should always be clean and free from dust, grease or loose, flaky material. If the surface has been gloss-painted, make sure the adhesive is compatible with the paint, ensure that the paint is not loose or flaking and then roughen its surface with sandpaper and wipe off the dust.
4. Make sure all surfaces are dry and protect the joint from moisture after closing until the adhesive has cured.
5. Use the correct tools to apply the adhesive (notched trowel with the right-sized notches) and make sure tools are cleaned after use, employing materials recommended by the adhesive manufacturer. Do not use solvents to clean adhesive from your skin.
6. Always observe the correct open time of a joint.
7. Do not attempt to make a glued joint when the temperature is outside the temperature range recommended by the adhesive manufacturer.
8. Do not store an adhesive beyond its recommended shelf life and always keep it within the storage temperatures recommended by the manufacturer.

Specific applications of adhesives

A number of specific applications of adhesives on site is listed in Table 7.1 together with the type of adhesive recommended for use.

Table 7.1 Adhesives recommended for particular applications

Application	Type of adhesive
Internal joinery	Casein, PVA (or the range of formaldehyde adhesives – not usually used on site)
Internal wall finishes	
Plasterboard	Gypsum plaster, PVA, gun-grade synthetic rubber
Decorative timber wallboard	Synthetic rubber (solvent based), synthetic rubber (emulsion)
Vinyl (rigid)	Synthetic rubber (emulsion), acrylic
Vinyl (flexible)	Acrylic
Textile	Synthetic rubber (solvent-based), PVA
Cork	Synthetic rubber (solvent-based), PVA
Ceramic wall tiles:	
to masonry base	Natural rubber latex cement, cement, PVA, synthetic rubber latex, epoxide, acrylic
to gypsum plaster	Natural rubber latex cement, PVA, synthetic rubber latex, acrylic
to timber-based boarding	Natural rubber latex cement, PVA, synthetic rubber latex, epoxide, acrylic
to paintwork	Natural rubber latex cement, PVA, synthetic rubber latex, acrylic
to metal	Natural rubber latex cement, PVA, synthetic rubber latex, epoxide, acrylic
Architraves/skirtings	Gun-grade rubber
Ceiling tiles	
Acoustic tiles	Gum spirit, synthetic rubber (solvent-based)
Polystyrene tiles	PVA
Gypsum plaster coving	Gypsum plaster
Insulating panels (glass fibre backed)	Synthetic rubber (solvent-based)

170

Floor finishes

Thermoplastic tiles — Bitumen, synthetic rubber (emulsion)

Vinyl asbestos tiles — Bitumen (not on mastic asphalt screed)

Flexible vinyl tiles
(BS 3261 Type B)

Flexible vinyl tiles
(BS 3261 Type A) — Bitumen, epoxide

Cushion-backed vinyl — Synthetic rubber (emulsion), acrylic, epoxide

Linoleum — Synthetic rubber (emulsion), acrylic, epoxide

Vinyl-backed linoleum — Gum spirit, lignin

Rubber — Synthetic rubber (emulsion)

Textile — Rubber solution, epoxide

PVC-backed textile — Synthetic rubber (emulsion), gum spirit

Cork (unbacked) — Synthetic rubber (emulsion), acrylic

Ceramic tiles: — Elastomer (solvent-based), synthetic rubber latex, gum spirit, lignin

to cementitious backings and
boards — Natural rubber latex cement, cement, synthetic resin emulsion, epoxide

to asphalt — Natural rubber latex cement

to vinyl — Cement

to steel — Natural rubber latex cement, epoxide

Wood blocks — Bitumen

External wall finishes

Ceramic wall tiles, mosaics, brick
slips, etc. — Cement-based adhesives, synthetic rubber latex emulsion with
cement/sand, epoxide

Decorative laminates to timber — Rubber solution (protecting joint from water penetration)

Joinery — Phenol formaldehyde, resorcinol formaldehyde (not usually site applied)

Additional explanatory detail can be found in the background section of this chapter.

Safety

All adhesive manufacturers now supply information on the safe handling of their products as required by the Health and Safety at Work Act, 1974. Particularly petroleum-based and synthetic-resin adhesives need careful handling and storage, not only on account of their flammability, but also because of the toxic fumes they give off. Adequate ventilation should always be ensured in areas where these adhesives are being used.

For further information see *Site Safety* by J. C. Laney in this Site Practice series.

Background

Principles of adhesion

If it were possible to produce two perfectly flat surfaces and bring them into contact, face-to-face, spontaneous adhesion would occur due to molecular forces of attraction at their interface. The nearest example of this phenomenon in everyday experience is the resistance that can be felt when two sheets of glass in face-to-face contact are separated.

But even the surface of glass does not have the degree of perfection required to produce spontaneous adhesion. The smoothest surface is rough by comparison with the molecular scale of smoothness required. As a result, a third component has to be introduced to overcome the surface irregularities of the parts to be joined together (the adherends) and form a rigid connection by solidifying or setting. This third component is a paste or liquid adhesive.

The degree of 'thickness' (or viscosity) of the adhesive will affect its gap-filling properties – its ability to fill the more major irregularities in the surface of the adherends. The adhesive must be compatible with both adherends; i.e. it must have molecular attraction for the materials of which they are made. For example, urea formaldehyde synthetic resin adhesive is not compatible with a rubber or polystyrene adherend. Urea formaldehyde is, however, compatible with wood and wood-based products. Table 7.2

Table 7.2 Material compatability groupings

Group A Polar group	Group B Non-polar group
Cellulose (wood, cotton, paper etc)	Rubber
Urea formaldehyde (UF)	Polystyrene
Phenol formaldehyde (PF)	Polyethylene
Water	Teflon
Alcohol	Benzene
Metal oxides	Mineral oils

All materials in each group are compatible with other members of the same group and incompatible with all members of the opposite group.

shows the two major compatibility groupings of adhesives and adherends.

Adhesives need to solidify in order to produce a firm joint. This setting process takes place as a result of one (or a mixture) of the following processes.

1. Through loss of solvent (water or organic solvent) due to evaporation. Non-aqueous solvent-based adhesives are often flammable. The setting of solvent-based adhesives can be reversed by reintroducing the solvent. This makes water-based adhesives, like animal and cellulose glues, vulnerable in damp conditions. Today there is a movement away from organic solvent-based adhesives towards water-based adhesives because of their greater ease of use and greater safety. Although organic solvent-based adhesives produce strong joints, they do release flammable and toxic vapours.

2. By the coalescing of binders in emulsion and latex adhesives as the water is absorbed in the adherends and evaporates.

3. By the cooling of an adhesive applied in a hot state. These hot-melt thermoplastic glues are usually supplied in a solid state and heated to 160–200 °C for application. They should not be confused with thermosetting adhesives which require heat to cure.

4. By a chemical cross-linking process. Synthetic resin adhesives, like resorcinol formaldehyde, melamine formaldehyde and the epoxide range belong to this chemically setting group. The process is irreversible and is

173

initiated by the introduction of a chemical called a hardener into the resin and/or the application of heat. Adhesives which set as a result of the application of heat are called thermosetting adhesives. They have little application on site due to the hot presses required for their use, but they are extensively used in the factory for the manufacture of timber components, plywood and other timber-board products.

What adhesive to use

For an adhesive to produce a permanent union between adherends it should:

1. Be compatible with both adherends.
2. As nearly as possible match the strength of the adherends.
3. Resist the conditions which it is likely to meet in use (dampness, heat, mould growth, chemical pollution, etc.), (see Table 7.3).
4. Fill the imperfections in the surfaces of the adherends to produce a consistent film; if necessary inert fillers such as wood flour may be used to produce gap-filling characteristics.
5. Remove minor surface impurities from the surface of the adherends.
6. In setting should not produce unacceptable stress due to shrinkage.
7. Have a setting time consistent with the nature of the work; i.e. sufficient pot life, sufficient open time of joint to allow correct assembly of its parts, and sufficient speed of setting to harden before the joint is likely to be disturbed.

Table 7.3 gives an indication of the resistance of various adhesives to degrading by moisture and their vulnerability to attack by bacteria. Table 7.4 lists many of the adhesives used in building in basic ingredient categories and explains their general characteristics and applications.

Table 7.3 Resistance classification of adhesives

| Type | Category | Durability | | | | Is attacked by bacteria |
		WBP	BR	MR	INT	
Starch	Natural				X	X
Plant–protein	Natural				X	X
Casein	Natural				X	X
Cellulose	Natural				X	X
Bitumen	Natural				X	
PVA	Thermoplastic				X	
Epoxide	Thermoset	X				
Urea formaldehyde	Thermoset			X		
Melamine formaldehyde	Thermoset		X			
Resorcinol formaldehyde	Thermoset	X				
Phenol formaldehyde	Thermoset	X				

WBP = weatherproof and boilproof: BR = boil-resistant (good weather resistance) MR = moisture resistant: INT = interior use only.

175

Table 7.4 Chief adhesives used in building-in ingredient classification

Group	Type	Constituents
Natural	Animal	Hides & bones
	Starch	Maize, cassova or wheat
	Plant-protein Cellulose	Soya bean
	Casein	Milk curd
	Gum spirit	Rosins in alcohol with fillers
	Lignin paste	Sulphite lye in water with fillers
	Bitumen	Bitumen
	Bitumen solution	Bitumen blends in solvents
	Bitumen/rubber	Aqueous emulsion of bitumen with natural or synthetic rubber latex
Elastomeric/ rubber	Rubber solution	Solution of natural or synthetic rubber in organic solvents, probably with resins and other modifiers

Description	Uses	Suitable bases made from:
Set is reversible with moist conditions. Little used in building today	Internal joinery, furniture	Wood
	Wallpaper	Portland cement, plaster, wood
	Internal joinery	Wood
	Expanded polystyrene, wallpaper	Portland cement, plaster, wood
Designated INT – interior glue. Sets part by evaporation; part by chemical action. Little resistance to moisture and micro-organisms	Internal joinery	Wood
Insoluble in water, but degrades under wet alkaline conditions. Careful priming needed when used with porous friable adherends	*Floors*: cork, lino, most PVCs, felt *Walls/ceilings*: wood fibre, expanded polystyrene, cork, polyurethane (unbacked)	Portland cement, wood, plaster, mastic asphalt
Softened by water. Used as a flooring adhesive in dry areas	*Floors*: Cork, lino, cork-backed PVC *Walls/ceilings*: lino, cork	Portland cement, wood
Applied hot; thermoplastic. No drying time needed	*Floors*: Wood block Bitumen roofing felt. Foam glass	Portland cement, wood
Solvent adhesive	*Floors*: Flexible PVC, vinyl asbestos, thermoplastic tiles	Portland cement, wood
Slow-setting	*Floors*: flexible PVC, vinyl asbestos, thermoplastic tiles, wood block, wood mosaic *Walls/ceilings*: cork, expanded polystyrene, polyurethane (unbacked), foamglass	Portland cement, wood
One-way wet-stick if one or both adherends is porous; two-way dry-stick contact adhesive with non-porous adherends. No positional adjustment possible after placing unless thixotropic grade used	*Floors*: cork, flexible unbacked PVC, rubber *Walls/ceilings*: wood fibre, ceramic, rubber, felt-backed PVC, chipboard, hardboard, decorative laminates, plasterboard, plywood, cork, felts, PVC or rubber-backed textiles,	Portland cement, wood, plaster

Table 7.4 Cont'd

Group	Type	Constituents
	Filled–rubber solution	Solution of synthetic or reclaim rubber containing fillers, resins and modifiers
	SBR emulsion	Aqueous dispersion of blends of styrene butadiene rubbers with synthetic resin modifiers and/or fillers
Thermoplastics	Acrylic emulsion	Emulsion of acrylate ester copolymers which may contain modifiers and/or fillers
	Polyvinyl acetate (PVA)	Aqueous emulsion of PVA with fillers
Thermosets	Epoxides (epoxy adhesives)	Wide range of chemically-setting synthetic adhesives

Description	Uses	Suitable bases made from:
	expanded polystyrene, paper-faced polyurethane, PVC skirtings & nosings	
Mastic-like consistency, applied with notched trowel or cartridge gun. Gap-filling ability up to 6 mm	*Walls/ceilings*: wood fibre, expanded polystyrene, rubber, chipboard, decorative laminates, plasterboard, plywood, plasticized PVC, wood wool slabs, backed and unbacked PVC, PVC or timber skirtings	Portland cement, wood, plaster
Good bond of moderate strength. Resistance to damp and alkaline conditions	*Floors*: cork, lino tiles, most PVCs, most textiles, thermoplastic tiles *Walls*: cork	Portland cement, wood, plaster; limited suitability for mastic asphalt
Resistant to damp conditions. Good bond under warm conditions	*Floors*: felt-backed lino tiles, most PVCs, most textiles *Walls/ceilings*: ceramic, lino, plasterboard, PVC paper-backed polyurethane	Portland cement, wood, limited suitability for mastic asphalt
One-part adhesive used primarily as a wood glue, but due to its elastic glue line, not used for structural joints. Slowly degraded by moist conditions. Designated INT-interior glue. *Two-part* modified adhesive can be used for external joinery; but not for structural joints	*Floors*: needle loom felt-backed PVC, felts *Walls/ceilings*: wood fibre, mineral fibre, expanded plastertiles, ceramic, expanded polystyrene, plasterboard, heavy wallpapers, paper-faced polyurethane	Portland cement, wood, plaster
Two or three-part adhesive in which liquid resin is mixed with hardener immediately before placing. High bond strength, good water and alkali resistance. Systems with fillers are used as adhesive mortar with negligable shrinkage but little flexibility	*Floors*: ceramic, flexible unbacked PVC *Walls/ceilings*: ceramic, metallic, chipboard, hardboard, plywood, rigid PVC *External*: brick slips, concrete repair, steel to concrete, stair nosings	Portland cement, wood, brick

Table 7.4 Cont'd

Group	Type	Constituents
	Formaldehyde	Range of chemically-setting synthetic resin adhesives
		Urea-formaldehyde (UF)
		Melamine/urea-formaldehyde (MF/UF) Melamine-formaldehyde (MF) Phenol-formaldehyde (PF) Resorcinol-formaldehyde (RF)
	Unsaturated polyester	Chemically-setting synthetic resin adhesive
Cement based	Polymer modified cement	Synthetic resin modified cement
	Sand/cement plus SBR	Sand/cement mortar modified with SBR rubber

Description	Uses	Suitable bases made from:
Two-part adhesives which can be of cold-setting type; but which set more quickly with heat Designated MR adhesive – moisture resistant moderately weather resistant – only limited life used externally.	Use generally restricted to timber-to-timber joints – mostly factory applied	
Designated BR adhesive – boil resistant – longer life externally, but not prolonged use Designated WBP adhesive – weather and boil proof – prolonged external use	Use generally restricted to timber-to-timber joints – mostly factory applied	
Two- or three-part with or without fillers, mixed immediately prior to application. High strength bonds, slight setting shrinkage. Good resistance to moisture and weathering. Systems incorporating fillers used as adhesive mortars	*Walls*: timber skirtings *External*: brick slips, concrete repair *Miscellaneous*: steel to concrete, stair nosings	Portland cement, brick
Grades available for wet and dry conditions	*Floors*: ceramic *Walls/ceilings*: mineral fibre, expanded plaster, ceramic, wood wool	Portland cement, brick *NOT* suitable for plaster and timber substrates
For external use, even to damp surfaces. Cheaper and easier to use than filled polyester and epoxide eqivalents. Has some initial grab	*External*: brick slips, concrete repairs	Portland cement, brick

Appendix 1

List of proprietary fixings in base material classifications

This list of mechanical fixing devices and manufacturers cannot aim to be comprehensive. It does, however, contain mention of the major manufacturers of proprietary fixing, all of whom have assisted in the preparation of this book. It does not endeavour to list manufacturers of standard non-proprietary items like nails or bolts.

Mass walling base

Plugs

Citmart: Syba anchor
Fischer: Nylon wallplugs; Wallplug Type S; Wallplug Type S-R; Wallplug S 10 RL; Wallplug S 10 H-R; Facade fixing; Frame fixing; Twist Lock Anchor Type GB
Hilti: HLD nylon anchor; FD nylon plug; HG anchor; HT frame anchor; Celanail
Mungo: Various patterns of nylon anchors
Rawlplug: Fibre plugs; Plastic plugs; Rawplastic compound; Rawlnut; Nailins
Spit: Plastic plug; Nylon anchor; Nail-in expansion anchor
Stainlessfix: UPAT Nylon wall plug; UPAT Nail plug UN; UPAT Nylon frame plug UR-S; UPAT Turbo low density anchor
Thorsman: Hiden anchor; Loden anchor; Nail plug TCP; Plug TP
Thunder Screw Anchors: Thuscan wallplug
Tucker: Molly anchor plugs; Molly brick fixing; Molly aerated concrete fixing; Molly collar plug; Molly frame fixing

Expanding anchors

Feb: Febolt
Fischer: Fischerbolts Type FA; Hammerset anchor Type EA; Heavy-duty nylon anchor; Heavy-duty steel anchor Type SL; Steel anchors Type MR; Suspension anchor Type L

182

Harris and Edgar: Ferrous and non-ferrous expanding anchors; Hemax 200 restraint fixing; Hemax 820 parallel expansion bolt; Hemax 2000 restraint fixing for thin claddings
Hilti: HA 8 R1 suspension anchor; HKD anchor; HSA anchor; HSL heavy duty anchor
Liebig: Ultra Plus expanding anchor; Safety bolt
Mungo: Various heavy duty anchors
Rawlplug: Rawlbolt loose bolt; Rawlbolt projecting bolt; Rawlbolt hook bolt; Rawlbolt eye bolt; Rawlok; Self-drill anchor; Duplex stud anchor
Spit: Heavy-duty double expansion anchor; Self-expanding stud anchor; Roc self-drilling anchor; Spit-fix anchor; Spit-grip Drop-in anchor; SDI Drop-in anchor; Frame anchor
Stainlessfix: UPAT Express anchor; UPAT expanding socket; UPAT PS anchor; UPAT wedge anchor
Thorsman: Torgrip
Tucker: Parabolt; Parabolt sleeve anchor

Insulation fasteners

Hilti: IN insulation nail; ID insulation nail
Stainlessfix: IPD insulation fastener; IPS insulation fastener

Cavity wall repair anchors

Harris and Edgar: Hemax 63

Chemical anchors

Cementation: Keyston chemical fixing
Fischer: Resin bonded anchor
Harris and Edgar: Resin anchor
Hilti: Resin anchor
Mungo: Chemical anchor
Rawlplug: Kenfix masonry anchor
Stainlessfix: UPAT chemical anchor; UPAT resin bonded Rockbolt
Tucker: Parabolt capsule anchor

Injection anchors

Fischer: Fischer injection system

Screw-in anchors

British Industrial Fasteners: Confas
Buildex: Tapcon

Masonry nails

Spit: Masonry nails
Rawlplug: Tufnails

Powder-actuated fasteners

Hilti: Various cartridge tools and pins
Rawlplug: Various cartridge tools and pins
Spit: Various cartridge tools and pins

Pneumatic-actuated fasteners

British Industrial Fastenings: Various tools and pins

Built-in fixings

Avon: Various ties, straps and joist hangers etc
Bat: Various ties, straps and joist hangers etc
Bevplate: Various ties, straps and joist hangers etc
Catnic: Straps, joist hangers etc
Halfen: Straps, corbels, anchors and ties etc
Harris and Edgar: Anchors, ties and corbels etc.
Hydro-Air: Straps and joist hangers etc.

Cast-in fixings

Avon: Various
Geo. Clark: Ties, sockets and corbels etc
Halfen: Corbels, anchors, channels and sockets etc
Harris and Edgar: Corbels, anchors, channels and sockets etc
Stainlessfix: Corbels, anchors, channels and sockets etc

Timber base

Most timber fixings, like nails, screws and bolt are common to several manufacturers. Exceptions are the Twinfast woodscrew and the Supadriv driving profile which are proprietary products of GKN.

Timber connectors

Bat: Toothed connector; Split ring connector; Shear plate connector

184

Other accessories

The following manufacturers make a variety of joist hangers, straps, nail plates and other accessories used in making timber-to-timber connections: *Avon; Bat; Bevplate; Catnic; Gangnail; Hydro-Air*

Metal base

Once more the simple metal-to-metal fixings like bolts, machine screws and set screws are common to a number of manufacturers. Only proprietary devices are listed here.

Base-deforming devices

Buildex: Teks screws; Standoff Teks; Dec-Loc; Scots; Twinseal cladding screw

British Screw Company: Sela all-metal fastener; Sela plastic headed fastener; Sela self-drilling fastener; Spacer screw

GKN: Self-tapping screws (various); Taptite

Linread-Fabco: Colorfixx cladding screw; Colorfixx roofing screw; Fab-Lok Pak Fix; Insul-Lok; Topseal B and AB; Topseal self-drilling

Powder- and pneumatic-actuated fasteners

Tools and pins manufactured by the following:
British Industrial Fastenings; Hilti; Rawlplug; Spit

Drilled-for fixings, rivets and self-tapping inserts

ISC: Clic rivet; Clufix; Kwik-sert; Press nut; Rifbolt
Linread-Fabco: Bulb-tite rivet

Thin-walled cellular bases

Fischer: Fixoplac-metal anchor; Hollow ceiling fixing Type K; Hollow door fixing Type A; Hollow wall fixing Type NA; KD spring toggle; ME brass rivet; NP nylon rivet; Wallplug Type SB

Rawlplug: Gravity toggle; Interset; Rawlnut; Spring toggle

Spit: CC high performance cavity anchor

Stainlessfix: UPAT Tric plug

Thorsman: Duo-Max TSP; Platti-Plug TPP

Thunder Screw Anchor: Rosegrip

Tucker: Molly blind screw anchor

Schedule of manufacturers' names and addresses

Avon Manufacturing (Warwick) Ltd, P.O. Box 42, Montague Road, Warwick CV34 5LS

BAT Building and Engineering Products, Haybrook, Halesfield Industrial Estate, Telford, Shropshire TF7 4LD

Bevplate Ltd, Rectory Farm Road, Sompting, Lancing, W. Sussex BN15 0DP

British Industrial Fastenings Ltd, BIF House, Gatehouse Road, Aylesbury, Bucks HP19 3DS

The British Screw Co Ltd, 153 Kirkstall Ropad, Leeds LS4 2AT

Buildex, ITW Ltd, Darville House, 4 Oxford Road East, Windsor, Berks SL4 5DR

Catnic Components Ltd, Pontygwindy Estate, Caerphilly, Mid Glamorgan CF8 2WI

Cementation Chemicals Ltd, Denham Way, Maple Cross, Rickmansworth, Herts WD3 2SW

Citmart, Ashford Airport, Hythe, Kent CT21 4LT

Geo. Clark (Sheffield) Ltd, Greasebrough Street, Rotherham, S. Yorks S60 1RG

FEB (Great Britain) Ltd, Albany House, Swinton Hall Road, Swinton, Manchester M27 1DT

Artur Fischer Ltd, Fischer House, 25 Newtown Road, Marlow, Bucks SL7 1JY

Gangnail Ltd, The Trading Estate, Farnham, Surrey GU9 9PQ

GKN Screws and Fasteners Ltd, P.O. Box 60, Heath Street, Smethick, Warley, W. Midlands B66 2SA

Halfen Ltd, Griffin Lane, Aylesbury, Bucks HP19 3AS

Harris and Edgar Ltd, Progree Works, 222 Purley Way, Croydon CR9 4JH

Hilti (Gt Britain) Ltd, Hilti House, Chester Road, Manchester M16 0GW

Hydro-Air International (UK) Ltd, Midland House, New Road, Halesowen, W. Midlands B63 3HY

ISC Fasteners, Datim Screw Co. Ltd, 180 Brooker Road, Waltham Abbey, Essex EN9 1JJ

Liebig Bolts Ltd, Silica Road, Amington Industrial Estate, Tamworth B77 4DT

Linread-Fabco Ltd, Swindon Road, Kingsditch Industrial Estate, Cheltenham, Glos GL51 9LT

Mungo Fasteners (UK) Ltd, Kirk-Sandall Industrial Estate, Doncaster, S. Yorks DN3 1QR

The Rawlplug Co Ltd, Rawlplug House, 147 London Road, Kingston-upon-Thames, Surrey KT2 6NR

Spit Fixing Ltd, Unit D2, Old Brighton Road, Lowfield Heath, Crawley, W. Sussex RH11 0QN

Stainlessfix Ltd, Unit 2A, Hawley Trading Estate, Hawley Lane, Farnborough, Hants GU14 8EH

Thorsman and Co (UK) Ltd, Thor House, Yarrow Mill, Chorley, Lancs
　　　PR6 0LP
Thunder Screw Anchors Ltd, Unit 11, Industrial Estate, Southwater,
　　　Horsham, W. Sussex RH13 7HQ
Tucker Fasteners Ltd, Construction Industry Division, 177 Walsall Road,
　　　Birmingham B42 1BP

Appendix 2

Glossary

adherend
> One part of a glued joint. A component which is connected to one or more components with an adhesive.

anchor
> A device used to make an anchorage fixing.

anchorage fixing
> A fixing device which makes a fixing to the base material by embedding in the material and not passing through it.

austempered steel
> A specially hardened steel involving treatment with heat and quenching.

base
> The major element of several parts, between which a fixing is made.

base-deforming fixing
> A fixing which is driven or otherwise forced into the base material and achieves its holding power by friction with the base, or by cutting matching threads in it.

bi-metallic corrosion
> Galvanic corrosion

blind fixing
> A device which is capable of making a connection between two or more elements of a joint without needing access to the further side of the joint.

cam out
> The action of a screwdriver, not completely perpendicular to a screw, flying out of the driving profile during screwing.

cavity fixing
> A device to make a fixing to a cellular base, utilizing the cavity inside the base and not requiring access to that cavity to set the device.

chemical fixing

A fixing device which depends on the setting of an adhesive to achieve its holding power (cf. mechanical fixing p. 190).

clearance hole

A pre-drilled hole which is of greater diameter than a mechanical fixing which passes through it; often formed in the subsidiary component.

component

The smaller part of a joint – the subsidiary component (cf. base p. 188).

compression

A force which tends to compress a material or fixing; for instance, a force parallel to the shank of a nail which presses it into the base.

contact adhesive

An adhesive with high grab which is applied to both adherends and allowed to dry before the joint is closed. After closing, no adjustment of the parts of the joint is possible.

contact corrosion

Galvanic corrosion.

drilled-for fixing

A fixing which requires the prior drilling of a hole before it can be set.

electrolytic attack

Galvanic corrosion.

embedment

In an anchorage fixing, the depth the fixing is set into the base material to achieve the required pull-out strength.

fired fixing

A fixing which is driven into the base material by the explosion of a cartridge or the release of compressed air within the fixing tool (see powder-actuated and pneumatic fasteners pp. 46 and 50).

first movement load

The point at which a fixing under load is first observed to move 0.2 mm.

fixing accessories

Pieces of metalwork which form an integral part of a fixing device but require connecting to the base by an anchorage device; e.g. an angle corbel (the accessory) connected by an expanding anchor to the mass walling base.

galvanic corrosion

Corrosion resulting from one metal having electrical contact with another metal.

galvanizing
> The coating of a ferrous element with zinc for corrosion protection by dipping the element in molten zinc.

gap-filling adhesive
> An adhesive specially formulated with a proportion of inert fillers to join together adherends with slightly rough surfaces.

grab
> The initial holding power of an adhesive when the joint is first closed.

gun-grouting
> The process of injecting grout (often resin-based grout) into a mortice to bed and set a cast-in fixing.

holding power
> The resistance of a fixing to pull-out failure.

imposed load
> All loads placed on a structure, other than those imposed by the weight of the parts of the structure; e.g. the wind pressure on a wall panel, but not the weight of the panel itself.

lead hole
> A hole of less diameter than the diameter of a screw (wood or self-tapping) into which it is driven (cf. clearance hole p. 189).

loadbearing fixing
> A fixing which carries the weight of the component, as well as maintains its correct location.

mechanical fixing
> A fixing that does not depend on the setting of an adhesive and is usually in the form of a metal (or occasionally plastic) device.

open time
> The length of time after the spreading of an adhesive before the joint is closed; usually stated by manufacturers as maximum and minimum periods between which the glue will have effective adhesion.

plug
> A screwable or nailable insert in mass walling.

popping
> The phenomenon of nails working out under conditions of use, particularly when intermittent loading causes timber structure to deflect.

pull-out failure
> Failure caused by an anchorage fixing, under load, pulling out of the base.

pull-out strength
> The resistance of a fixing to pull-out failure.

pull-through failure
>   Failure caused by the subsidiary component, under load, being
>   forced over the head of the fixing device.

relieved shank
>   Where the diameter of the shank of a bolt or screw is less than
>   the crest of the thread.

restraint fixing
>   A fixing which maintains a component in its correct location,
>   but is not required to carry its weight.

safety factor
>   The amount by which the ultimate load of a fixing is reduced to
>   establish a safe working load (cf. ultimate load, p. 192 first
>   movement load p. 189).

self-weight
>   The weight of the component without including any imposed
>   load.

shank
>   The cylindrical part of a bolt or screw.

shear
>   A force acting at right angles to the axis of the fixing.

shelf life
>   The period during which an adhesive can be stored in its sealed
>   container without danger of deterioration. ˙

sherardizing
>   The diffusion of hot zinc dust on to a ferrous element to give it
>   added corrosion resistance.

spalling
>   The fracture of mass walling, resulting in flaking and breaking
>   away, around the point of entry of the fixing device.

subsidiary component
>   The minor part of an assembly, connected to the base by
>   mechanical or chemical fixing.

substrate
>   In adhesive terminology, the major part of a connection – the
>   base.

s.w.g.
>   Standard wire gauge. A method of measuring wire diameter,
>   used to describe nail thickness; now rapidly being superseded by
>   metric dimensions.

tension
>   A force applied to a fixing parallel to its axis and acting to pull
>   the fixing away from the base.

thermoplastic adhesive
> An adhesive which softens with heat.

thermoset adhesive
> An adhesive which sets due to a chemical reaction brought about by heat.

thixotropic
> A thixotropic material is one whose normal consistency is that of a thick paste, which thins on stirring.

thread
> A spiral groove formed in or on the shank of a screw or bolt. The root of the thread is at the base of the groove; the crest at the external angle or apex between adjacent grooves.

through-fixing
> A fixing that is made by a mechanical device, such as a bolt, passing right through the joint, both subsidiary component and base.

ultimate load
> The load at which a fixing fails.

unrelieved shank
> The shank of a screw or bolt in which the diameter of the shank is the same as the diameter of the crest of the thread.

waisted bolt
> A bolt having its shank diameter reduced below the root diameter of the thread.

wet-stick adhesive
> An adhesive which is closed while the adhesive is wet, as opposed to a contact adhesive which is closed after the adhesive has dried to a degree.

working load
> The ultimate load of a fixing divided by a safety factor.

zinc plating
> The electroplating of a ferrous metal device with zinc for corrosion protection.

Appendix 3

Comparative diameters of various devices

(reprinted from *Mechanical Fixing Devices in the Building Industry* by Paul Marsh and Derrick Beckett, Construction Press 1975)

Diameter (mm)	Bolts and Machine Screws					Wood screws gauge	Nails (s.w.g.)	Self-tapping screws (A & B gauge)
	ISO Metric	BA	BSW	BSF	UNC/UNF			
							20 . 0.90	
1.0							19 . 1.00	
							18 . 1.20	
							17 . 1.40	
						0 . 1.52		
	M1.6						16 . 1.60	
						1 . 1.78		
							15 . 1.80	
2.0	M2						14 . 2.00	
						2 . 2.08		
		8 . 2.18						2 . 2.18
							13 . 2.30	
						3 . 2.39		
		7 . 2.50						
							12 . 2.60	
		6 . 2.74				4 . 2.74		
					4 . 2.84			4 . 2.84
							11 . 2.90	
3.0						5 . 3.10		
	M3	5 . 3.20	1/8 . 3.20					
							10 . 3.30	
						6 . 3.48		
					6 . 3.50			6 . 3.50
		4 . 3.51						
						7 . 3.70	9 . 3.70	
								7 . 3.83
4.0	M4		5/32 . 4.00					
		3 . 4.08						
							8 . 4.10	
					8 . 4.16			8 . 4.16
						8 . 4.17		
						9 . 4.50	7 . 4.50	
		2 . 4.75						
			3/16 . 4.80	3/16 . 4.80				
					10 . 4.83			10 . 4.83
						10 . 4.88	6 . 4.88	
5.0	M5							
		1 . 5.31						
							5 . 5.40	
								12 . 5.49
						12 . 5.59		
							3 . 5.60	
							3 . 5.90	

mm		Gauge / Metric dia.	Imperial dia. / Metric dia.	Imperial dia. / Metric dia.	Imperial dia. / Metric dia.	Gauge– / dia. in mm	SWG– / dia. in mm	Gauge– / dia. in mm
6.0	· M6	0 · 6.00	1/4 · 6.40	1/4 · 6.40	1/4 · 6.40	1/4 · 6.30	2 · 6.30	14 · 6.15
7.0						16 · 6.94	1 · 6.70	
8.0	· M8		5/16 · 7.90	5/16 · 7.90	5/16 · 7.90	18 · 7.72	0 · 7.00	
9.0						20 · 8.43		
10.0	· M10		3/8 · 9.50	3/8 · 9.50	3/8 · 9.50			
11.0	· M12		7/16 · 11.1	7/16 · 11.1	7/16 · 11.1			
12.0			1/2 · 12.7	1/2 · 12.7	1/2 · 12.7			
13.0								
14.0			9/16 · 14.3	9/16 · 14.3	9/16 · 14.3			
15.0	· M16		5/8 · 15.9	5/8 · 15.9	5/8 · 15.9			
16.0								
17.0								
18.0								
19.0	· M20		3/4 · 19.1	3/4 · 19.1	3/4 · 19.1			
20.0								
21.0								
22.0			7/8 · 22.3	7/8 · 22.3	7/8 · 22.3			

Index